m-Business: The strategic implications of wireless technologies

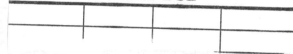
DATE DUE

m-Business: The strategic implications of wireless technologies

Stuart J. Barnes

Amsterdam • Boston • Heidelberg • London • New York • Oxford
Paris • San Diego • San Francisco • Singapore • Sydney • Tokyo

ELSEVIER
BUTTERWORTH
HEINEMANN

Butterworth-Heinemann
An imprint of Elsevier
Linacre House, Jordan Hill, Oxford OX2 8DP
200 Wheeler Road, Burlington MA 01803

First published 2003

British Library Cataloguing in Publication Data
A catalogue record for this book is available from the British Library

Library of Congress Cataloguing in Publication Data
A catalogue record for this book is available from the Library of Congress

ISBN 0 7506 5623 9

For information on all Butterworth-Heinemann publications
visit our website at www.bh.com

Typeset by Newgen Imaging Systems (P) Ltd, Chennai, India
Printed and bound in Great Britain by Biddles Ltd, *www.biddles.co.uk*

Contents

List of figures

List of tables

Introduction

It is a common euphemism to say that the Internet and related technologies are changing the ways we live. Clearly, these technologies will affect peoples' lives in ways that have yet to be imagined. Indeed, if the Internet pundits are correct, few areas of our lives will remain untouched (e.g. Negroponte, 1995; Tapscott, 1997). The Internet has proven to be an easy and efficient way of delivering a wide variety of services to millions of 'wired' users; as of September 2002, the estimated number of Internet users stood at 605.6 million (Nua, 2003a), rising to 1 billion by 2005 (IDC Research, 2001). Users can enjoy convenient global access to services via the Web browser; for the supplier, server-based service development via off-the-shelf tools can be very quick and convenient, providing a short time-to-market (724 Solutions, 2003).

One of the most significant changes promises to be in the way business is conducted. From being primarily a resource for the rapid and secure communications of the scientific and military communities, the Internet and related technologies are developing into the communication systems of choice for a variety of business activities in a diverse range of industries (Nua, 2003b).

Throughout the 1990s, another technology that has played an increasingly important role in society is the mobile phone. Again, this is a technology in an age where time is short and the weight attached to convenience is high. From a penetration of only 8 per cent in 1995, some 73 per cent of the UK adult population owned a mobile phone in 2002 (OFTEL, 2002). Similar patterns can also be seen in Japan, Hong Kong, Finland, New Zealand, the United States, and many other countries. In some areas, such as some parts of Scandinavia and South-East Asia, the saturation of mobile phone ownership is now in excess of 80 per cent (Rauhala, 2002), often with penetration rates of more than 95 per cent in the young adult market. In June 2002, the number of mobile phone users worldwide reached 1 billion for the first time (eMarketer, 2002).

Of late, the impact of the Internet and wireless telecommunications has taken a new turn. Until recently, these technologies have followed very separate paths. However, since the late 1990s, mobile technology providers have been working to bring convergence, enabling a variety of wireless data communication technologies such as the wireless Internet. The products of this partnership are sophisticated wireless data services, centering on mobile data access and electronic messaging on mobile devices. The market for these services is diverse, and the most commonly cited applications are in the business-to-consumer (B2C) and business-to-employee (B2E) segments. Such applications are built on some fundamental value propositions, such as ubiquitous access to information, the personal nature of devices and customization, and contextual properties of the device and user, such as time, location, preferences, and the task at hand. In the consumer space, the wireless applications have included person-to-person messaging, e-mail, banking, news, games, music, shopping, ticketing, and information feeds. In the business space, applications include sales force automation, navigation, tracking, field force automation, wireless telemetry, and the mobile office.

Moreover, the commercial value of this new revenue stream is predicted to be very significant. Mobile telephony offers the potential platform for unprecedented penetration of the Internet and services such as mobile (m-) commerce. Simply speaking, m-commerce is defined as any transaction with a monetary value – either direct or indirect – that is conducted over a wireless telecommunication network. Worldwide, m-commerce revenues (including peer-to-peer payments) are expected to exceed $25 billion by 2006 (Frost and Sullivan, 2002).

More broadly speaking, mobile (m-) business is likely to have a tremendous impact on organizations, as wireless technologies and applications begin to challenge the existing processes, strategies, structures, roles of individuals, and even cultures of organizations. Here, m-business is defined as the use of the wireless Internet and other mobile information technologies for organizational communication and coordination, and the management of the firm. Indeed, by 2004, cost savings could permit wireless business services around the world to generate annual value of up to $80 billion, and at least as much value could be created if corporations used wireless services to improve their current offerings or to deliver new ones (Alanen and Autio, 2003).

Features of the book

This book explores in depth the strategic impact of wireless technologies driving the emergent concept of m-business. This text has arisen from extensive investigation into the impacts of wireless technologies in a variety of areas of business and organization, each highly dependent upon recent technological developments. It has also arisen from a personal review of the available professional and academic literature on this and related topics, based on experience, and in the context of recent developments in the field.

While the book will be certainly be accessible and of interest to managers who are examining the strategic issues associated with wireless technologies, applications, and services for the first time, its primary audiences are senior undergraduates and Master's students studying for business-related degrees. Students who are about to embark on research in this emergent area should also find the book of particular help in providing a rich source of contemporary material reflecting leading-edge thinking.

The chapters of this book illustrate the wide array of business opportunities afforded by m-business. They describe and discuss the important strategic, managerial, and technological issues that follow in the wake of an organization deciding to embrace wireless technologies. Chapters have been created to bring a balance of conceptual and practical issues, focusing on recent and emerging trends. Where possible, the book examines wireless issues from an international perspective, pointing to specific examples from around the globe.

It is, of course, impossible to cover all aspects of this emerging topic. The focus of this book has been on attempting to cover a selection of the core, recent, or possibly more important areas of m-business, with reference to different markets, technology foundations, applications, services, and impacts for organizations. The implications are that whilst technological aspects are covered in some detail, this is always in a mode accessible to the manager. This is, after all, an introductory text aimed at management students.

Structure of the book

This book's 10 chapters are structured into three sections, each emphasizing different but interrelated aspects of the m-business landscape.

Section one: strategic overview of m-business markets

The first section provides an introduction to the book as a whole, explaining core background material on the nature of m-business and its relationship to different markets. Specifically, the section explores the two principal markets for m-business – the consumer and the enterprise. In doing so, it examines the technologies, services, and applications underpinning offerings in these areas. Each of the chapters in this section provides a framework to assist in understanding the nature and implications of 'm-' for the particular market.

Section two: strategic wireless technologies

The focus of the second section is on exploring in greater detail a range of wireless technologies underlying the development of m-business applications and services. While section one provides a brief overview of the technological landscape, this section highlights four key areas of wireless that have profound strategic implications for m-business services. In particular, it examines the nature and implications of the highly successful iMode service in Japan, the wireless application protocol service platform, short-range wireless technologies, and location-based services.

Section three: emerging m-business applications

The final section of the book explores a variety of m-business applications in greater detail. Clearly, there is a vast array of possible applications that we could examine in this last part of the book. Therefore, only a limited number of applications are selected. The first chapter in this section, 'Enterprise mobility: concept and cases', provides an insight into the types of applications impacting on wireless in the organization. It provides a background, framework, and case studies exploring the impact of wireless on the enterprise. The remaining three chapters focus largely on mobile applications aimed at the consumer. Each of these applications has been selected to provide a different perspective on wireless consumer services. Specifically, the application areas are wireless advertising, banking, and news services.

Chapter summary

In order to further elaborate on the content of the book and the range of issues covered, each part will now be described in somewhat greater detail.

As mentioned above, Section One, *Overview of wireless markets*, provides a strategic overview of the consumer and business markets for wireless services. Focusing on business-to-consumer markets, chapter 1 aims to examine how value is added in the stream of activities involved in providing m-commerce to the consumer. As such, it analyses the key players and technologies that form part of the m-commerce value chain, providing a foundation for future strategic analysis of the industry later in the book. The chapter concludes by drawing on some of the key factors that may influence the take-up of m-commerce – including technological and other issues.

Much of the literature on applications of the wireless Internet has predominantly focused on business-to-consumer markets, following the patterns in the media and e-commerce research. Notwithstanding this, it is now becoming clear that mobile networking will provide a tremendous impetus to the development of other strategic applications for businesses. Chapter 2 explores how wireless network computing will create value in the business sector. The chapter draws on a number of very recent and forthcoming wireless applications for organizational use and concludes with some reflections on the future of m-business applications in the firm.

Section Two, *Strategic wireless technologies*, examines in detail four key areas of wireless technology that have strategic implications for business. Chapter 3, 'Experiences with Japan's iMode service', explores a key exemplar of wireless services in consumer markets. In most cases the adoption of wireless Internet technology by consumers has been slow, typified by the lukewarm reception accorded to wireless application protocol in Europe and the United States. However, in Japan, the number of wireless Internet subscribers has grown phenomenally, driven by the mobile Internet services provided by NTT DoCoMo. This chapter focuses on wireless Internet services in Japan and, in particular, the iMode phenomenon. It provides a detailed background on iMode, as well as some discussion of a number of competing services. Further, using technology acceptance theory, it examines the success of iMode in Japan, and the implications for the international wireless Internet market. It also draws on recent market experiences with next-generation technologies and international iMode services.

Outside Japan, the development of wireless Internet services has followed a different path. Under the present technological constraints of low bandwidths and high latency in wireless networks, as well as the low power and small screens of handheld devices, a key standard has emerged for Internet service provision – the Wireless Application Protocol (WAP). For most industrialized markets, it is WAP that has provided the means for bringing the Internet and a range of services to the wireless consumer. As such, WAP has created a whole new set of dynamics in the wireless industry driven by the possibilities accorded to value-added service provision. This chapter explores the strategic implications of WAP service provision, drawing on a number of analytical frameworks. The purpose of the chapter is to provide a deeper understanding of the forces impacting on the ability of WAP to succeed in Internet service provision. It concludes with a discussion of the future of WAP services and some practical implications.

One important facet of wireless communications to emerge in the last few years is the high potential of short-range wireless connectivity for close proximity interaction and communication. Such technology removes the tremendous constraints provided by wired connectivity. Typically, personal computers and related devices are connected with special cables, whilst wireless devices such as mobile phones use proprietary networks to communicate. Technological developments have offered some new directions to such problems of connectivity by providing a convergence between wireless and computing technologies. Standards, such as Bluetooth and WiFi, promise to allow unlimited connectivity between devices by embedding short-range wireless transceivers 'under the skin' of products. The commercial potential of these technologies is enormous, ranging from the wireless office or home, through to in-vehicle connectivity and location-based advertising. This chapter examines the range of technologies available for short-range embedded wireless interactivity. It explores some of the key areas of application of such devices, including in the home, the workplace, in-transit, and for public spaces. The chapter also evaluates the key benefits and problems associated with these applications. It concludes with reflections on the future penetration of such technologies.

The final chapter in the section, 'Location positioning technologies and services', examines a key area of current m-commerce development. Indeed, location-based services (LBS) are heralded as the next major class of value-added services that mobile network operators can offer their customers. Using a range of network- and handset-based positioning techniques, operators will be able to offer entirely new services and improvements on current ones. Popular examples cited include emergency caller location, people or asset tracking, navigation, location-based information, or geographically sensitive billing. The purpose of this chapter is to examine the technologies, applications, and strategic issues associated with the commercialization of LBS. The chapter concludes with some predictions on the role of LBS in m-commerce.

Section Three, *Emerging wireless applications*, explores a number of areas of m-business development. Clearly, there is a vast array of applications that could be investigated here. Therefore, the emphasis is on the examination of a specific subset of applications, each selected to fulfil a particular role or provide a particular perspective. Chapter 7, 'Enterprise mobility: concept and potential', provides a detailed investigation of wireless applications in the organization. Mobile networking will provide a tremendous impetus to the development of a variety of strategic enterprise applications. This chapter explores the emerging area of wireless applications in the business. It provides a background to conceptual ideas of enterprise mobility, and a framework for understanding the development of enterprise mobility in organizations. The chapter also begins to apply these conceptual ideas in a number of original case studies. The chapter rounds off with a summary and some conclusions regarding the future of enterprise mobility.

The remaining three chapters focus largely on wireless applications for consumer markets. Chapter 8 examines wireless advertising, predicted to be a major growth area for marketing. The personal, always-at-hand nature of devices, along with the ability to assess context-dependence (e.g. time and location), presents a rich platform for wireless advertising. Using a range of wireless platforms, operators will be able to offer very different advertising services that go beyond those of the wired Internet. The chapter begins by reviewing the core technology platforms for wireless advertising, and an examination of the various types of emergent wireless advertising services. The chapter continues by providing a research model for wireless advertising, touching on wireless ad processing, structures, and outcomes. The chapter concludes with some important research questions and predictions on the future role of wireless advertising.

Using a variety of platforms, services are being created to enable mobile devices to perform many activities of the traditional Internet, albeit in a reduced format for mobile devices. Chapter 9 focuses on one of these areas – m-banking – which is also one of the first areas of commercial transaction on the wireless Internet. Banking has extended in many different ways in recent years, including telephone and online banking. m-Banking provides yet another channel for banking services, and in emerging markets, provides some possibility for becoming a primary channel. This chapter examines the nature of m-banking services, the strategic implications of m-banking, and strategic positioning of m-banking services in different markets. The chapter concludes with a discussion of the future for m-banking services.

The final chapter of the book examines one of the first areas of information service provision on the wireless Internet – news services. It investigates user perceptions of wireless Internet services in general, and news services in particular. The chapter reports on the evaluation of wireless Internet news sites using the WebQual/m instrument. From initial application in the domain of traditional Internet Web sites, the instrument has been adapted for sites delivered using WAP. The WebQual approach is to assess the Web-site quality from the perspective of the 'voice of the customer', a method adopted in quality function deployment. The WebQual/m instrument is used to assess customer perceptions of information, site- and user-oriented qualities. In particular, the qualities of three UK-based WAP news sites are assessed via an online questionnaire. The results are reported and analysed and demonstrate considerable variations in the offerings of the news sites. The findings and their implications for mobile commerce are discussed and some conclusions and directions for further research are provided.

As you will now be aware, m-business is a complex and diverse subject. It is not simply concerned with technological issues, but incorporates aspects of strategic management, marketing, operations management, and behavioural science, among others. Such an interdisciplinary perspective is critical if the subject domains are to be understood fully. Recent examples of m-business offerings that overestimate technology and underestimate consumers exemplify this point. WAP offerings for complex browsed content over second-generation technology has received a poor market perception in all but a few markets; simple messaging, based on the short message service (SMS), on the other hand, has been a phenomenal success. For this reason, as you have seen, we advocate a broader management viewpoint. The issues debated here are far too important to be left to the technologists; although technology is an important enabler, the vision, strategy, and management of the transition to bold new m-business models lies squarely in the hands of managers. To reap the real rewards of m-business, management competence is crucial.

We hope you find this book of interest and that it raises some important issues relevant to consideration in your study, research, or organizational context. As you do so, we all take one more step in the wireless world.

References

Alanen, J. and Autio, E. (2003). Mobile business services: a strategic perspective. In B.E. Mennecke and T.J. Strader, eds, *Mobile Commerce: Technology, Theory and Applications*. Hershey: Idea Group Publishing, pp. 162–184.

eMarketer (2002). One billion mobile users by end of Q2. http://www.nua.ie/surveys/index.cgi?f=VS&art_id=905357779&rel=true, accessed 28 April 2002.

Frost and Sullivan (2002). Mcommerce transactions to hit USD25 billion. http://www.nua.ie/surveys/index.cgi?f=VS&art_id=905357769&rel=true, accessed 29 April 2002.

IDC Research (2001). A billion users will drive e-commerce. http://www.nua.ie/surveys/index.cgi?f=VS&art_id=905356808&rel=true, accessed 28 May 2001.

Negroponte, N. (1995). *Being Digital*. New York: Alfred Knopf.

Nua (2003a). How many online? http://www.nua.com/surveys/
how_many_online/index.html, accessed 12 February 2003.

Nua (2003b). Communications usage trend survey. http://www.nua.ie/,
accessed 31 January 2003.

OFTEL (2002). Key trends in fixed and mobile telephony, and
Internet. http://www.oftel.gov.uk/publications/research/2002/
trenr0602.pdf, accessed 20 December 2002.

Rauhala, P. (2002). Comparison between Finnish and Japanese
mobile markets. In *Proceedings of the International Workshop on
Wireless Technology and Strategy in the Enterprise*, Berkeley,
California, 15–16 October.

724 Solutions (2003). *Commerce Goes Mobile*. San Francisco: 724
Solutions Inc.

Tapscott, D. (1997). *Growing Up Digital*. New York: McGraw-Hill.

Overview of wireless markets

Chapter 1

The mobile commerce value chain in consumer markets

Introduction

Barely before Internet-facilitated e-commerce has begun to take hold, a new wave of technology-driven commerce has started – m-commerce. Fuelled by the increasing saturation of mobile technology, such as phones and personal digital assistants (PDAs), m-commerce promises to inject considerable change into the way certain activities are conducted. Equipped with micro-browsers and other mobile applications, the new range of mobile technologies offer the Internet 'in your pocket' for which the consumer possibilities are endless, including banking, booking or buying tickets, shopping, and real-time news. Focusing on business-to-consumer markets, this chapter aims to examine how value is added in the stream of activities involved in providing m-commerce to the consumer. As such, it analyses the key players and technologies that form part of the m-commerce value chain, providing a foundation for future strategic analysis of the industry. Drawing on some of the key factors that may influence the take-up of m-commerce – including technological and other issues – the chapter also provides predictions regarding the future of m-commerce.

The m-commerce value chain

Well before the introduction of m-commerce, the rapid adoption of personal computers (PCs) and greater accessibility of Internet infrastructure had already fuelled immense growth in the digitization of products and services. In this new digital economy, consumer online services demand that diverse inputs must be combined to create and deliver value. No single industry alone has what it takes to establish the online digital economy; success requires inputs from diverse industries that have only been peripherally related in the past (Schleuter and Shaw, 1997; Tapscott, 1995). As a result, cooperation, collaboration, and consolidation have been the key watchwords, as arrangements are struck between companies in complementary industries. Noticeably, companies in telecommunications, computer hardware and software, entertainment, creative content, news distribution, and financial services have seized opportunities by aligning competencies and assets via mergers and acquisitions, resulting in a major consolidation of information-based industries (Symonds, 1999).

The challenge in this section is to move beyond an understanding of traditional Internet-based commercial activities to the relatively new and unexplored area of m-commerce. Like any product or service, m-commerce involves a number of players in a chain of value-adding activities that terminates with the customer. Whilst traditional value chain analysis (Porter and Millar, 1985) could be used to unravel this

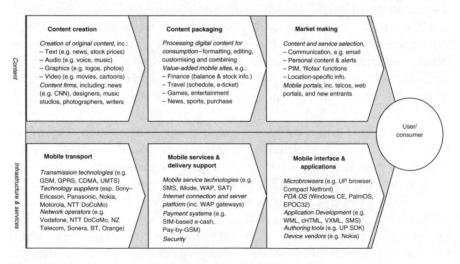

Figure 1.1   The m-commerce value chain

complexity, there are a number of more advanced value chain techniques specifically aimed at the new media that are preferable (Loebbecke, 2001). In particular, the European Commission (1996) developed a framework for new media publishing that has been productively employed in other areas of online activity (Loebbecke, 2001; Schleuter and Shaw, 1997). Within this chapter, the framework is adapted for m-commerce and used to analyse the players, technologies, and activities involved in providing m-commerce – as summarized in Figure 1.1.

The basic model consists of six core processes in two main areas: (a) content and (b) infrastructure and services. Let us examine each of these, in turn.

Infrastructure and services

This part of the value chain is associated with providing the technological platform for services, particularly the development and standards creation for hardware, software, and communications.

Mobile transport

This is the basic network involved in communications, including transportation, transmission, and switching for voice and data. This includes major telecommunications players such as AT&T, NTT DoCoMo, and Vodafone, and current high-speed transmission technologies such as the European third generation (3G) Universal Mobile Telecommunications System (UMTS) and 2.5G General Packet Radio Service (GPRS).

Infrastructure equipment vendors and network operators are the main players involved in adding value in the transport element of the framework. The leading suppliers for mobile network infrastructure equipment – which in Europe include Sony–Ericsson, Siemens, Nokia, Motorola, and Lucent – have developed and continue to develop mobile data solutions. The innovative capabilities of these companies are driving the next wave of technological developments. Table 1.1 summarizes some of the key present and future transport technologies. A key distinction between network standards is the access method they use (as shown in Figure 1.2). Traditionally, second generation (2G) standards such as Global System for Mobile Communication (GSM) and Personal Communications Services (PCS) have been based on Time Division Multiple Access (TDMA).

Table 1.1 Key mobile network technologies

Standard	Description	Max. speed[a]
GSM (Global System for Mobile Communication)	The prevailing mobile standard in Europe and most of the Asia-Pacific region – around half of the world's mobile phone users	14.4 Kbit/s
PCS (Personal Communications Services)	A standard based on TDMA, used particularly in the United States, central/south America and other countries	14.4 Kbit/s
PDC (Personal Digital Cellular)	A standard used in Japan. Uses packet-data overlay on second-generation networks to achieve 'always-on' data communication and a higher speed	28.8 Kbit/s
HSCSD (High Speed Circuit Switched Data)	A circuit switched protocol based on GSM. It is able to transmit data at around four times the speed of GSM by using four radio channels simultaneously	57.6 Kbit/s
GPRS (General Packet Radio Service)	A packet switched wireless protocol as defined in the GSM standard offering instant, 'always on' access to data networks	115 Kbit/s (burst)
EDGE (Enhanced Data rates for Global Evolution)	This is a higher bandwidth version of GPRS and an evolution of GSM. EDGE conveniently provides a migration path to UMTS by implementing necessary modulation changes	384 Kbit/s
IMT2000 (International Mobile Telecommunications)	This is a third-generation standard. Three rival protocols have been developed: Universal Mobile Telephone System (UMTS) in Europe, CDMA 2000 in the United States, and Wideband-CDMA in Japan. The development of the standard requires significant investment in infrastructure	2 Mbit/s

Note: [a]Speed is measured in kilobits (1 kilobit = 1024 bits) and megabits (1 megabit = 1,048,576 bits) per second (Kbits/s and Mbits/s, respectively)

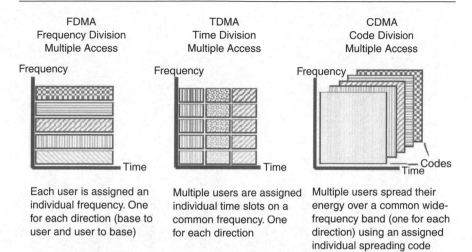

Figure 1.2 Network access standards explained (Epstein, 2002)

More recently, although some pockets of 2G Code Division Multiple Access (CDMA) have existed in the United States and South Korea for some time, 3G standards have introduced CDMA to the global market, introducing a much higher level of bandwidth efficiency.

Mobile network operators – such as Sonera, AT&T, NTT DoCoMo, and Vodafone – are also an important part of the transport process. In addition, these players are in a strong position to leverage their infrastructure advantages in transport to enable movement along the value chain towards mobile services, delivery support, and market making. Typically, these operators control the billing relationship and/or SIM (subscriber identification module) card on the phone and, as we have seen in the consumer marketspace, are ideally positioned to become Internet service providers (ISPs) or portals, thereby establishing a transport pipeline for content services.

Mobile services and delivery support

An important aspect of m-commerce provision is the availability of platforms for delivering services. This space involves, for example, the infrastructure in connecting to the Internet, security, the server platform, and payment systems. Standards such as the WAP, SMS, and iMode are key building blocks towards enabling the delivery of Internet services via mobile handsets. The establishment of these platforms goes hand-in-hand with network standards and key interest groups ensure this, including the UMTS Forum, International Telecommnications Union (ITU), ETSI (European Telecommunications

Table 1.2 Key mobile service technologies

Service	Description
SMS (Short Message Service)	Allows text messages of up to 160 characters to be sent to and from mobile handsets via a store-and-forward system. Although a large proportion of this is based on person-to-person communication and voicemail, other services such as news, stock prices, and SMS chat are growing in popularity. Around 500 billion messages were sent in 2001
MMS (Multimedia Message Service)	This is a new messaging service supporting graphics and audio currently emerging in key mobile markets. It plans to build on the success of SMS
CB (Cell Broadcast)	Not to be confused with citizen's band (CB) radio; this is another text messaging service. However, unlike SMS, CB provides a one-to-many broadcast facility that is ideal for push-based information services such as news feeds
SAT (SIM Application Toolkit)	This allows applications to be sent via CB or SMS in order to update SIM cards, e.g. for downloading ringing tones. Data security and integrity are standard features making it a popular choice for mobile banking. The WAP 2.0 standard is compatible with SAT
WAP (Wireless Application Protocol)	WAP is a universal standard for bringing Internet-based content and advanced value-added services to wireless devices such as phones and PDAs. In order to integrate as seamlessly as possible with the Web, WAP sites are hosted on Web servers and use the same transmission protocol as Web sites, that is, hypertext transfer protocol (HTTP). The most important difference between Web and WAP sites is the application environment. Whereas a Web site is coded mainly using hypertext markup language (HTML), WAP sites use Wireless Markup Language (WML), based on eXtensible Markup Language (XML)

MExE (Mobile Station Application Execution Environment)	This standard is aimed at incorporating Java into the mobile phone and providing full application programming. MExE is compatible with WAP but incorporates many other sophisticated services including voice recognition and positioning technology
J2ME (Java 2 Micro Edition)	A version of the Java language designed for small devices. This is somewhat similar to MExE
iMode (information mode)	iMode uses a variant of HTML for the provision of Web pages. iMode-enabled Web sites utilize pages that are written in compact HTML (cHTML) – a subset of HTML 4.0 designed with regard to the restrictions of the wireless infrastructure
iAppli (information application)	From January 2001, an upgraded version of iMode became available in Japan to premium customers. The new service, called iAppli (short for 'information applications'), is based on Java. Applications can be downloaded and stored, thereby eliminating the need to continually connect to a Web site. Further, constantly changing information is automatically updated at set times, e.g. stock prices or weather forecasts
PDA Web Clipping	This technology allows popular PDA devices, such as Palm and Handspring, to access dynamic and updated HTML content via a modem. Web clipping is used in combination with applications stored on the device
PDA Syncing	This allows PDAs to store or cache content without the use of a wireless modem. Content is updated when users synchronize ('sync') or connect their PDA to the Internet via computer connection

Standards Institute), GAA (GPRS Applications Alliance), WAP Forum, and others. Some of the current wave of mobile service technologies is given in Table 1.2.

Middleware infrastructure is crucial to driving applications. Typically, this involves hardware and software to link the wireless and wired networks. In the WAP environment, for example, gateways are required – either at the mobile operator's or corporate customer's site (see Chapter 4). The nature of the service is also quite different to the standard Web; while standard Web content is received via the PC browser, mobile content, based on the appropriate service protocol, is received via a mobile device 'micro-browser'.

Other important aspects of mobile services and delivery support include payment systems and security – both of which are inextricably linked in m-commerce.

The development of security standards has been a key area for the attention of standards groups. For consumer-browsed content, the main service available outside of Asia has been WAP. In WAP 1.x, security is handled using Wireless Transport Layer Security (WTLS), which encrypts data based on Netscape's Secure Sockets Layer (SSL) technology. WAP 1.3 introduced Public Key Infrastructure, which uses digital certificates, certificate authorities, strong asymmetric encryption, and digital signatures to ensure integrity, privacy, authenticity, and non-repudiation. WTLS allows for the negotiation of encryption parameters between the client and server, thus ensuring a secure channel for communication. Although WTLS does not provide end-to-end security, the chances of a problem are small because hackers can breach security only when sensitive data passes through the WAP gateway (Boncella, 2002). WAP 2.0, in addition to extending the bearer services to enable GPRS and 3G, provides a more efficient and secure method of interaction with wired Web servers. Since WAP 2.0 includes a version of TLS (Transport Layer Security) in its WAP Device stack, security is improved since the WAP proxy no longer needs to translate the WTLS protocol into the TLS protocol when sending data to a wired Web server, and vice versa. Overall, WAP 2.0 provides better end-to-end security (Nichols and Lekkas, 2002).

In addition to the traditional monthly billing model, various e-cash payment methods have been piloted and used based on cash stored (via SIM or credit-sized card) and transferred via the wireless network or at point-of-sale using short-range wireless connectivity. In Finland, payment via mobile phone is particularly popular;

Figure 1.3 demonstrates a variety of mobile payment scenarios facilitated via the operator Sonera. However, it is clear that mobile cash and payment systems still have some way to go towards maturity. In the United Kingdom, Visa piloted a debit smartcard system called Visa Cash in 1999, while France Telecom launched a similar service called Iti Achat. Such a system relied on a dual-slot phone such as Motorola's StarTac D model that can accept a credit-sized card. However, the extra size and weight of devices favoured a different approach. For example, in Finland, Sonera's Pay-by-GSM enables the user to dial a number to receive a charge to a prepaid phone or for a deduction from a mobile account.

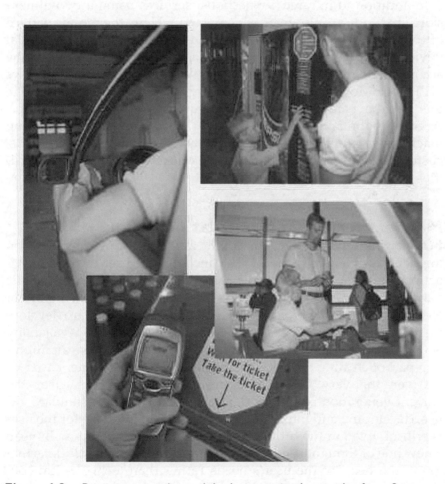

Figure 1.3 Payment using the mobile device – visual examples from Sonera

Another method, trialled by Visa, Nokia, and Merita-Nordbanken, uses the dual SIM concept, where a second SIM is a Visa credit, debit, and bankcard. KDDI/Au is currently piloting a similar system for its 3G service in Japan, using its 'next-generation UIM (user identity module)'. The card can record private information, such as credit card details, and data can be communicated via infrared. A similar system is about to be launched in Japan in mid-2003, when IYBank launches its Mobile Cash Card service via iMode. The service allows cash card data to be loaded directly onto the cell phone (NE Asia, 2003). Cash data is exchanged with an automated teller machine (ATM) via infrared. Subsequently, the phone can be used in the same way as a conventional cash card to transfer cash via infrared communications.

In addition to infrared connections, services using a two-dimensional bar code are also increasing in Japan. Here, the mosaic pattern of the bar code is displayed on the cell-phone screen, and a special device reads the pattern to identify a customer and to make the account settlement. The most popular service – 'Combien?' – allows NTT DoCoMo customers to receive a bar code bill via the cell phone that can be paid at appropriately equipped convenience shops. Another popular service – 'C-mode' – uses Coca-Cola vending machines to purchase a variety of items. In addition to soft drinks, customers can buy other items such as tickets for amusement facilities, ring tones, and graphics.

Mobile interface and applications

In the mobile environment, developing and integrating an application interface for the user is critical. This includes the user interface, navigation, and application/middleware development, as well as the authoring tools. For example, micro-browsers have been an important element of providing browsed content, along with software development on service platforms such as WAP, iMode, and SMS. Availability and penetration of appropriate handset technology is also an important part of creating the right environment for the customer base.

Given the very different nature of communication over the current generation of mobile devices as compared to standard PC use, developing and integrating an application interface for the user is critical. Even in the future, the very nature of mobility will mean a new line of thought is needed for developing mobile solutions that get to the heart of the user's needs rather than technological constraints. Some of the important players in this value-adding space include technology platform vendors, application developers, and mobile device vendors.

Technology platform vendors provide the operating systems (OS) and micro-browsers for mobile devices. Micro-browsers perform the same function as those of the Web, such as Netscape and Internet Explorer, but have been designed explicitly for the mobile environment. They have reduced functionality tailored to the present mobile devices. Phone.com's browser dominates the WAP market, and they have support from all but two major phone manufacturers (see www.phone.com); Ericsson and Nokia have developed their own micro-browsers. On the iMode platform, Compact NetFront is the most popular micro-browser, used in 75 per cent of devices.

The OS market for PDAs has been dominated by Microsoft (Pocket PC, Microsoft CE), Symbian (EPOC32), and 3Com (PalmOS). A large number of applications are built for the offline palmtop/ PDA environment. Connectivity is being improved by the development of applications on these platforms, but this is somewhat overshadowed by developments in the more lucrative smartphone market via platforms such as iMode, WAP, and SMS, where large numbers of interactive, consumer-oriented applications have been developed (see below). Whereas iMode uses a variant of hypertext markup language (HTML) for service provision, and more recently Java has become used (e.g. in iAppli), WAP adopts Wireless Markup Language (WML) as the format for displaying Web pages over the WAP phone (based on XML – eXtensible Markup Language – a meta-language used to describe the content and format of data). In most cases, HTML content needs to be rewritten for WML. A voice recognition language for the mobile environment, Voice XML – a natural choice of interface for a voice-oriented device – is under development (see www.vxmlforum.org).

In the smartphone market, as in the PDA market, the brand and model are the most important part of the purchase decision; the service provider or network provider is less important (Peter D. Hart Research Associates, 2000). One of the reasons for this is the importance of 'image' and 'personality' to young customers – as associated with specific mobile phones. In the mobile interface and applications component of the value chain, this places a lot of power in the hands of smartphone producers, who also decide which technologies are incorporated into the end products. These producers must continue to innovate and support leading edge technologies and services in their new products if m-commerce is to prosper. A selection of recent handheld devices, in a number of device

Figure 1.4   Recent handheld devices – in various device categories: (a) 2002/3 phones (which include features such as colour screens, poly sound, built-in cameras, clip-on accessories, and Java); (b) 2002/3 integrated voice and data devices (combining the capabilities of a phone and PDA); (c) 2002/3 PDAs (with built-in short range wireless connectivity and/or PC, compact flash card, or cable capability for wireless data communications)

categories, is provided in Figure 1.4. Recent innovations emerging in the market have included 3G phones, devices combining the capability of a mobile phone and PDA, video capture and playback, high-quality audio playback (e.g. MP3), and short-range wireless connectivity. Although all of these devices are oriented to mobile data communications, their user orientations differ significantly, trading off reach and richness to target a specific segment of the mobile market. Figure 1.5 depicts some of the variation in the specific types of devices, noting some use characteristics in the continuum of choices.

Content

This portion of the m-commerce value chain is associated with creating, transforming, and delivering to the consumer the content for mobile Internet services.

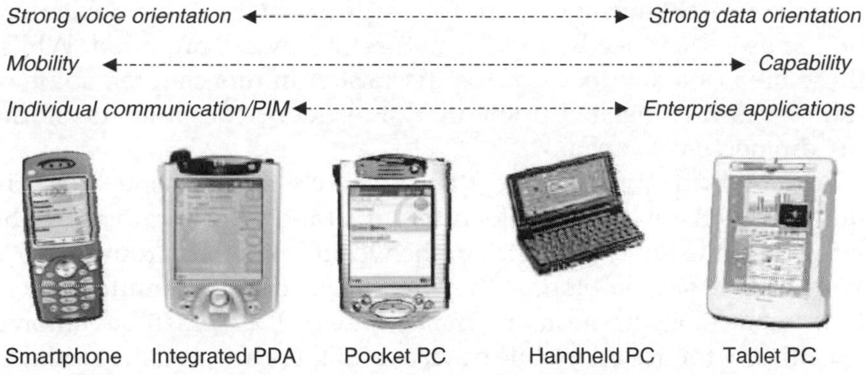

Strong voice orientation ◄----------------------------------► Strong data orientation

Mobility ◄--► Capability

Individual communication/PIM ◄----------------------------► Enterprise applications

Smartphone   Integrated PDA   Pocket PC   Handheld PC   Tablet PC

Figure 1.5   The spectrum of user device options for mobility

Content creation

The concept of digital content creation and delivery is the same in m-commerce as in e-commerce (Choi *et al.*, 1997), although the specific format will differ due to the nature of mobile devices (Mennecke and Strader, 2003). Typically, on the wireless Internet, such content will include:

audio, for example, voice, ringing tones, and MP3 music files;
graphics, for example, photos, logos, and picture messages;
video, for example, movies, animations, and personal recordings;
text, for example, news, stock prices, film listings, advertisements, product descriptions, and restaurant locations.

Such content can be easily modified, consumed repetitively by the same or different users, and is fast and cheap to reproduce; respectively, online delivered content (ODC) has the fundamental attributes of transmutability, indestructibility/non-subtractivity, and reproducibility (Loebbecke, 2001).

The creation of digital content for delivery via the mobile Internet raises some important issues (see Choi *et al.*, 1997; Loebbecke, 2001). These include: interactivity and customization (dynamic vs. generalized content); time-dependence (e.g. real-time stock quotes vs. a dictionary); intensity of use (used once, several, or many times); operational format (executable vs. fixed document); pricing (e.g. based on usage, subscription, or commission); and externalities (who gains and loses from consumption, e.g. positive externalities from positive reviews of content and negative externalities from illegal copyright infringement). To a large extent, these are dependent on the user, their needs, and the nature of services provided. Whilst these are not a key focus of the discussion in this chapter, some of the issues are touched-upon in downstream components of the m-commerce value chain.

Although plentiful on the standard Web, digital content is still quite limited on the mobile Internet. iMode, for example – the largest mobile Internet service in the world – only has around 50,000 Web sites. A key problem is the effort needed to transmute content for mobile consumption; in many cases ODC must be tailored specifically for use on mobile devices and there are certain standards to enable this (such as WML, SMS, cHTML, and others – see above).

A variety of forward-thinking companies have positioned themselves for the provision of mobile services, often as part of a portfolio of delivery channels, which, in addition to traditional channels, might encompass mobile Internet, wired Internet, and digital TV. For example, major news providers such as CNN, Reuters, and Webraska (a traffic news company) are just some of the companies leading the way by positioning their products using a variety of distribution channels. For example, Reuters is providing selected content to key mobile portals (e.g. Yahoo and Excite), network providers (e.g. Nokia and Ericsson), and mobile content packagers (e.g. the Financial Times), as well as developing its own mobile portals for a broad range of news markets. In this way, Reuters is beginning to branch out further downstream in the mobile commerce value chain – moving into the market making value space in Figure 1.1.

Content packaging

In most cases, digital content must be transmuted, edited, customized, or combined to provide consumable content for the user. Firms in the content packaging stage of the value chain focus on aggregating and transforming information for distribution to wireless devices. Here, value is added by the reconfiguration of data into the most appropriate package for user consumption. This could include, for example, packaging Reuters' stock information for FT.com, the online version of the business and financial newspaper. The information on this site is also customisable for individual users who may, for example, only be interested in certain financial markets.

There are a plethora of mobile sites situated in this space, including: sports (e.g. news, information, and sports services), online games, financial services (e.g. mobile banking and stock price information), entertainment (e.g. movie listings and schedules, restaurant information, streaming video, graphics, ringing tones, and comics), mobile newspapers, payment (e.g. parking meters, vending machines, and photo booths), information (e.g. telephone directory, message boards, and products and services), and travel (e.g. travel information, flight timetables, and e-check-in). Figure 1.6 provides some examples of services. Figure 1.6(a) shows an electronic postcard service and Figure 1.6(b) shows a mobile game available from Sonera in Finland. Figure 1.6(c) illustrates a stock information service

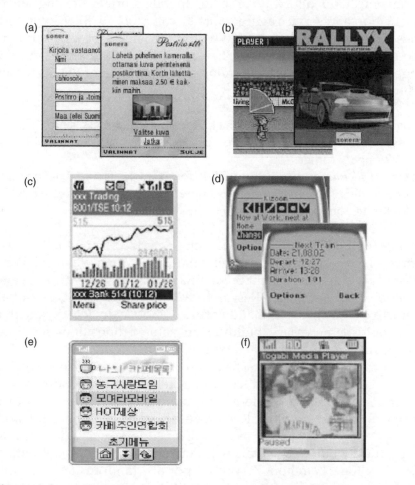

Figure 1.6 m-Commerce services – some examples: (a) postcard service; (b) games; (c) stock information; (d) travel information; (e) Café (message board, SMS, chat); (f) streaming video

available from iMode in Japan, whilst Figure 1.6(d) demonstrates the Kizoom travel service available in the United Kingdom. Finally, Figure 1.6(e) demonstrates a mobile café service and Figure 1.6(f) shows a screenshot of streaming video available from KT Freetel in South Korea.

Market making

The key business-to-consumer market makers on the mobile Internet are mobile portals (or m-portals) – revenues of which are predicted to be $42 billion by 2005 (Ovum, 2000). Literally, the word 'portal' means a doorway or gate; mobile portals are high-level information and service aggregators (Ticoll *et al.*, 1998) or intermediaries (Chircu and Kauffman, 2001) that provide a powerful role in access to the mobile Internet. Their main aim is the provision of a range of content and services tailored to the needs of the customer, including:

> *communication*, for example, e-mail, voice mail, and messaging;
> *personalised content* and *alerts*, for example, news, stock alerts, betting, sports, and weather;
> *personal information management* (PIM), for example, 'filofax' functions; and
> *location-specific information*, for example, traffic reports, nearest ATM, film listings, hotels, and restaurant bookings.

As such, mobile portals are usually characterized by a much greater degree of customization and personalization than standard Web-based portals in order to suit the habits of the consumer (Clarke and Flaherty, 2003; Durlacher Research, 2000). The compact, personal, interactive, and always-at-hand nature of mobile devices dictate that this should be necessary, building on value propositions for time, location, and personal characteristics.

In Europe, portal players have attempted to build on existing brands, competencies, and customer relationships to develop a subscriber base. By 2001, there were more than 200 WAP portals in Europe (Bughin *et al.*, 2001). Key players have been mobile operators (e.g. Sonera's Zed portal), technology vendors (e.g. Ericsson and Palm), traditional Web portals (e.g. Yahoo! and AOL), retail outlets (e.g. Carphone Warehouse's mviva portal), random new entrants (e.g. banks and media companies), and new independents. In contrast, the Japanese portal market is generally dominated by the operators – NTT DoCoMo, KDDI/Au, and J-Phone/Vodafone. It is also worth noting that in the European model, the provision of content is much looser than that of Japan, where the content service must answer to a need of the mobile user (characterized by the *nami-densetsu* or 'wave-legend'), not just simply bring information. In Europe, content services are more fragmented and may simply be brought to a new media without focusing on the usage of that media and to the specific needs of the user.

Now that we have some appreciation of the m-commerce value chain, the next section takes a critical perspective on the future possibilities of this phenomenon. Here, the issue is whether the value chain can be further leveraged to create a profitable partnership with consumers.

The future potential of m-commerce

Modern wireless communications represent the convergence of two key technology trends of the 1990s: portability and networking. Combined with the diffusion of mobile telephones and the Internet, the potential of these developments for commercial mobile services is enormous. However, although there are some high-profile examples of success in certain markets, mobile commerce has not, as yet, become well established. Typically, the problems hindering the progress of m-commerce have been the high cost, slow transmission rates, high power consumption of devices, inadequate mobile interfaces, and lack of interest or perceived need (Goodman, 2000; Mennecke and Strader, 2003). Nevertheless, it is interesting to examine whether such barriers are being overcome, thereby unlocking the future potential of m-commerce.

Advances in transmission networks

Current 2G networks, such as GSM and PCS, have limited speeds for data transmission. These circuit-switched networks require the user to dial-up for a data connection. The present wave of network investment is providing faster, packet-switched networks – such as GPRS and CDMA 2000 1x – that deliver data directly to handsets, which are, in essence, always connected. These are more suitable for *ad hoc* m-commerce, such as instant messaging or alerts, as well as reducing the cost of data transmission for the consumer. In the future, full implementation of 3G networks promise yet higher transmission speeds and more sophisticated m-commerce interaction, although based on a much bigger investment in infrastructure. Nevertheless, even when fully implemented and when coverage improves the use of 'rate adaptation techniques' means that in most mobile conditions the rate of 3G transmission will be less than the maximum speed, and is typically expected to be around 384 Kbit/s as adapted to the prevailing environmental conditions.

Although recent advances in network technologies have helped to alleviate the problem of transmission speed, this is only part of the story. Problems of high power consumption and cost remain important barriers to adoption: where high transmission speeds can be achieved power consumption of devices is also high, and the lofty initial outlays in infrastructure and frequency licensing in countries such as the United Kingdom and Germany mean that initial charges for services are also likely to be high. In order to provide a stronger platform for future service development, the next generation of mobile devices must address power consumption issues as a priority. Furthermore, operators need to think very carefully about consumer value and pricing models for the next generation of services to encourage adoption. KDDI in Japan, for example, seem to have struck the right balance, and already have 3 million subscribers for is 'au' branded 3G network, launched in April 2002 (Openwave, 2002). Interestingly, KDDI's promotional pricing for the service has captured a very high subscription among the Japanese student population.

The potential of service platforms

At present, WAP adoption by consumers is both patchy and limited (see Chapter 4). According to Raczkowski (2002), although the number of WAP phones is growing in developed telecommunications markets, only a fraction have so far used mobile phones to make purchases; citing a recent AT Kearney Mobinet survey, she suggests that 16 per cent of users in the combined markets of Finland, France, Germany, Italy, Japan, Spain, the United Kingdom, and the United States have a Web-enabled phone, yet less than 1 per cent of respondents have used them to make purchases. Thus far, the promise of m-commerce has not been embedded in the buying habits of consumers. Intention to purchase via WAP phones in the survey stood at 12 per cent, being highest in Japan (17 per cent) and lowest in the United States (3 per cent). The low adoption rates point to an important issue – consumers cannot predict usage of 'high-potential' services that do not yet exist.

Clearly, large numbers of consumers are not convinced of the benefits and advantages of mobile commerce technology compared to traditional channels. In most developed markets, the expectations of consumers have not been met and WAP has been considerably over-sold. In all, 26 per cent of consumers in the eight countries surveyed by AT Kearney cited a lack of interest or perceived need as the single largest reason for not intending to purchase products and services through a mobile device (Raczkowski, 2002). Another survey found problems in terms of security, high cost (until all networks become packet-switched and the pricing model changes), and limited infrastructure (from networks and devices) (Barnes *et al.*, 2001). A key issue is the limitation of technology and the non-subtractive nature of services; WAP is not a replacement for the wired Internet and involves an important trade-off between richness and reach in providing data services (Wurster and Evans, 2000).

In the present environment, SMS has proved the preferred service for data communication on mobile phones in most Western countries (Frost and Sullivan, 2003). From a market perspective, an analysis of SMS usage has also shown unparalleled access to the age group from 15 to 24 years – a group that has proved extremely difficult to reach with other e-commerce media (Pura, 2003). Key reasons for the overwhelming success of SMS include privacy, flexibility, absence of face-to-face contact, and easy availability of the SMS medium, which is now widespread on phones (Puca, 2001). In addition to person-to-person communication, SMS has also been used for commercial services. Examples of m-commerce services via SMS are many and various and include: share price alerts, ads, news, bulletin services for nightclubs, dating services, eBay auction alerts, bank statements, and football results services. Advertising is likely to be a high potential area (see Chapter 8). For example, a recent, large-scale study in the United Kingdom found that SMS advertising from trusted sources (such as a carrier) was as acceptable as TV or radio advertising (Enpocket, 2002). Another study of 200 SMS campaigns found that virality and response were also high, with 23 per cent of people showing or forwarding an advertising message to a friend, and

46 per cent of individuals responding to the best performing campaigns, typically by replying, visiting a Web site, visiting a store, or buying a product (Enpocket, 2003). Nonetheless, SMS is very limited in its capability and does not allow the flexibility of browsing Web content, typically operating effectively alongside other media. Use of SMS for m-commerce is not widespread. The successor to SMS – MMS – will provide a richer environment for messaging, integrating graphics and audio, but the browsing limitation still holds.

Aside from SMS and WAP, considerable attention is now being drawn to the HTML-based iMode standard in Japan, discussed in Chapter 3. Launched in February 1999, iMode now has more than 34 million subscribers, two-fifths of the Japanese mobile Internet population (Mobile Media Japan, 2003). This is more than three times that of the competing WAP service, EZWeb. The m-commerce model used for iMode has proved extremely successful and profitable for NTT DoCoMo, the owners of the iMode brand and service. Analysts put this success down to a number of reasons including technological investment, market dominance, vertical integration in technology development, and the low penetration of expensive wired Internet. NTT DoCoMo have so far unsuccessfully tried to 'export' this model to other markets via a partnering strategy. In some senses the Japanese iMode example is somewhat unique and perhaps unlikely to be emulated in Europe or the United States. However, there are some clear lessons for other wireless markets. IMode is a brand standing for key concepts like simplicity, functionality, and meeting consumer needs (WireFree-Solutions, 2000). WAP, on the other hand, is merely a bundle of technologies and protocols, which on its own does not deliver value to the end-user.

Other possible service platform alternatives include standards based on Java – a 'write once, run anywhere' programming language – to provide a full-application execution environment (see above). These standards are primarily aimed at 3G networks and the next generation of powerful smartphones. In South Korea and Japan, 3G phone users are already successfully using Java phones. In South Korea, by May 2002, there were approximately 5 million downloads per month of Java applications on the Qualcomm's Binary Runtime Environment for Wireless (BREW) platform – offered by KT Freetel – from the then 650,000 users of Java phones (Epstein, 2002). In Japan, there were approximately 140,000 users of the iAppli service in September 2002 (Shimbun, 2002).

The potential of location-specific applications

Although platforms such as WAP provide some potential for moulding applications to mobile lifestyles, this is, at present, quite limited. Indicative research suggests that only a limited number of services are truly suited to WAP, with the suitability to task of services such as news and booking a flight being far in excess of share prices and e-mail (Barnes *et al.*, 2001; *PC Magazine*, 2000). As iMode has demonstrated, if m-commerce is going to succeed, clearly it needs to adapt to users rather than current technological constraints.

In search of true mobility, numerous technologies are beginning to be used to expand the opportunities afforded to mobile devices by enabling location-specific m-commerce, often referred to as location (l-) or positioning (p-) commerce. The applications of these technologies are endless and many examples have been cited.

One development is the use of wireless in personal area networks (PANs) to enable m-commerce. Standards such as Bluetooth and IEEE 802.11 allow a new wave of short-range device interactivity and provide cheap, low-power, high-data-rate connectivity for portable devices in a limited area (see Chapter 5). Here, for example, the roaming phone user can be provided with information, alerts, or even advertisements based on local interaction with PANs. Customers could also conduct transactions with their mobile phones at the point of sale – vending machines and ATMs being the best-known examples. In a more complex scenario, the customer looking for a specific product or price could conceivably scan or enter a code into a phone; when the customer walks past a store with the right product or price the phone could send an alert. Stores could even send advertising alerts in an effort to tempt customers inside.

In addition to technologies that allow localised communication and interaction by *being* at a certain location, there are others that work by *knowing* the location of a mobile device. These have created the platform for a plethora of services in areas such as safety, navigation and tracking, information, and location-based transactions (see chapter 6). Of these, safety is the key market driver, driven by policy mandates in the United States and Europe. The commercial potential of this is enormous. Advertising, roadside assistance, fleet management, road pricing, and location-based products are some of the other possible LBS under development. Figure 1.7 provides two examples of consumer LBS: a service finder from Finland (Figure 1.7a)

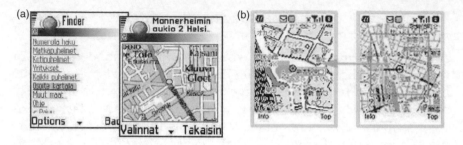

Figure 1.7 Consumer applications of LBS – some examples: (a) service finder (Finland); (b) navigation aid (Japan)

and a navigation aid from Japan (Figure 1.7b). Indeed, the Strategis Group (2000) estimates that LBS could be worth $3.9 billion by 2004. In terms of business-to-consumer m-commerce, one of the most basic LBS being offered by mobile operators is the mobile Yellow Pages. In this type of service, the roaming user asks the question: 'what's near me?' For example, items such as locations of restaurants, shops, public transport, or nearby ATMs may be useful to the user as they move through an unfamiliar city. Weather or traffic information can also prove useful; Bell Mobility's Book4golf service allows the user to locate a North American golf course, book a tee time, and get a location-specific weather forecast.

Conclusions

The convergence of wireless telecommunications and the Internet provides many exciting possibilities and predictions for the growth of mobile commerce. This chapter has attempted to examine how value is added in this new era of mobile Internet. It outlines some of the key players and technologies that underpin the delivery of m-commerce services. The result is a detailed value chain analysis that provides a solid foundation for understanding the concept of m-commerce in consumer markets.

Alongside, the chapter also provides some predictions regarding the future potential of m-commerce. Without doubt, the emergence of standards such as WAP, iMode, GPRS, and 3G will drive the wireless Internet forward. However, in many cases the technologies and services currently on offer to consumers do not reflect well on the possibilities achievable in the mobile medium. A fresh, creative look at the needs of the wireless consumer will shed new light on this issue and possibly break some of the technological constraints that are holding m-commerce back. Location-specific technologies could present some important pieces of this puzzle and enable LBS applications that get to the heart of adding value in a mobile environment. Similarly, short-range wireless technologies will allow a new era of localized wireless interactivity for mobile data appliances. The next few years will be very important in the development of technologies and services that provide truly mobile commerce.

Acknowledgements

An earlier version of this chapter first appeared as: Barnes, S. (2002). The mobile commerce value chain: analysis and future developments. *International Journal of Information Management*, **22**, 91–108.

References

Barnes, S., Liu, K., and Vidgen, R. (2001). Evaluating WAP news sites: the WebQual/m approach. *Proceedings of the European Conference on Information Systems*, Bled, Slovenia, June.

Boncella, R.J. (2002). Wireless security: an overview. *Communications of the AIS*, **9**, 269–282.

Bughin, J., Lind, F., Stenius, P., and Wilshire, M. (2001). Mobile portals: mobilize for scale. *The McKinsey Quarterly*, No. 2, 118–127.

Chircu, A.M. and Kauffman, R.J. (2001). Digital intermediation in electronic commerce – the eBay model. In S. Barnes and B. Hunt, eds, *Electronic Commerce and Virtual Business*. Oxford: Butterworth-Heinemann, pp. 45–66.

Choi, S.-Y., Stahl, D., and Whinston, A. (1997). *The Economics of Electronic Commerce*. New York: Macmillan.

Clarke, I. and Flaherty, T.B. (2003). Mobile portals: the development of m-commerce gateways. In B.E. Mennecke and T. Strader, eds, *Mobile Commerce: Technology, Theory and Applications*. Hershey: Idea Group Publishing, pp. 185–201.

Durlacher Research (2000). *Internet Portals*. London: Durlacher Research.

Enpocket (2002). *Consumer Preferences for SMS Marketing in the UK*. London: Enpocket.

Enpocket (2003). *The Response Performance of SMS Advertising*. London: Enpocket.

Epstein, M. (2002). CDMA technology keynote – Qualcomm. *Proceedings of Telecom NZ Wireless Data Symposium 2003*, Wellington, August.

European Commission (1996). *Strategic Developments for the European Publishing Industry Towards the Year 2000 – Europe's Multimedia Challenge*. DG XIII/E. Brussels: European Commission.

Frost and Sullivan (2003). *World Mobile Commerce Markets*. London: Frost and Sullivan Ltd.

Goodman, D.J. (2000). The wireless Internet: promises and challenges. *IEEE Computer*, **33**, 36–41.

Loebbecke, C. (2001). Online delivered content: concept and potential. In S. Barnes and B. Hunt, eds, *Electronic Commerce and Virtual Business*. Oxford: Butterworth-Heinemann, pp. 23–44.

Mennecke, B.E. and Strader, T., eds (2003). *Mobile Commerce: Technology, Theory and Applications*. Hershey: Idea Group Publishing.

Mobile Media Japan (2003). Japanese mobile Net users. http://www.mobilemediajapan.com/, accessed 15 January 2003.

NE Asia (2003). Cellular phones may replace wallets. http://neasia.nikkeibp.com/wcs/leaf?CID=onair/asabt/news/208033, accessed 24 September 2002.

Nichols, R.K. and Lekkas, P.C. (2002). *Wireless Security: Models, Threats, and Solutions*. New York: McGraw-Hill.

Openwave (2002). Openwave supports growth to three million KDDI CDMA 2000 1x mobile subscribers with proven WAP 2.0 technology. http://www.openwave.com/newsroom/2002/20021112_kddi_3million_1112.html, accessed 12 January 2003.

Ovum (2000). Wireless portal revenues to top USD42 Bn by 2005. http://www.nua.ie/surveys/index.cgi?f=VS&art_id=905355888&rel=true, accessed 5 July 2000.

PC Magazine (2000). Usability labs report – WAP services. September, 134–147.

Peter D. Hart Research Associates (2000). *The Wireless Marketplace in 2000*. Washington DC: Peter D. Hart.

Porter, M. and Millar, V.E. (1985). How information gives you competitive advantage. *Harvard Business Review*, **63**, 149–160.

Puca (2001). Booty call: how marketers can cross into wireless space. http://www.puca.ie/puc_0305.html, accessed 28 May 2001.

Pura, M. (2003). Case study: the role of advertising in building a brand. In B.E. Mennecke and T. Strader, eds, *Mobile Commerce: Technology, Theory and Applications*. Hershey: Idea Group Publishing, pp. 291–308.

Raczkowski (2002). *Mobile Ecommerce: Focusing on the Future.* New York: Dash30.

Schleuter, C. and Shaw, M.J. (1997). A strategic framework for developing electronic commerce. *IEEE Internet Computing*, **1**, 20–28.

Shimbun, N.K. (2002). DoCoMo stumbles at home, abroad. http:// neasia.nikkeibp.com/wcs/leaf?CID=onair/asabt/news/210825, accessed 15 October 2002.

Strategis Group (2000). European wireless location services: strategies and outlook. http://www.strategisgroup.com/press/pub/ wlocate.htm, accessed 1 December 2000.

Symonds, M. (1999). Business and the Internet: survey. *Economist*, June 26, 1–44.

Tapscott, D. (1995). *The Digital Economy*. New York: McGraw-Hill.

Ticoll, D., Lowy, A., and Kalakota, R. (1998). Joined at the bit – the emergence of the e-business community. In D. Tapscott, A. Lowy, and D. Ticoll, eds, *Blueprint to the Digital Economy*. New York: McGraw-Hill.

WireFree-Solutions (2000). *WAP vs. I-Mode – Let Battle Commence.* Madrid: WireFree-Solutions.

Wurster, T. and Evans, P. (2000). *Blown to Bits*. Boston: Harvard University Press.

Chapter 2

Wireless applications in the firm's value chain

Introduction

Although the literature on commercial wireless applications has predominantly focused on business-to-consumer markets, following the patterns in the media and e-commerce research, it is now becoming clear that mobile networking will provide a tremendous impetus to the development of other strategic applications for businesses. It is now becoming clear that the impact of mobile computing and m-commerce goes much further; wireless technologies have the potential to transform activities both within and between businesses (Alanen and Autio, 2003). Industry sources predict that corporate demand is likely to drive the wireless market forward and many applications are being developed for wireless enterprise computing (Wrolstad, 2002). Indeed, by 2004, cost savings could permit wireless business services around the world to generate an annual value of up to $80 billion, and at least as much value could be created if corporations used wireless services to improve their current offerings or to deliver new ones (Autio *et al.*, 2001).

This chapter picks up on this important issue by exploring the potential impact of mobile applications on the traditional value chain of the business. It provides an examination of the types of applications that are being developed, piloted, or used in a variety of organizations from around the globe. The chapter continues by providing an assessment of the key benefits of mobile business applications in the value chain, and some of the likely impacts. It concludes with a summary of the chapter and some reflections on the future of wireless applications in the firm.

The impact of mobile applications on the firm's value chain

An effective way of investigating the potential opportunities of wireless information technology is through a systematic analysis of a company's value chain – the series of interdependent activities that bring a product or service to the customer (Porter, 1980; Porter and Millar, 1985). In different settings, information technology (IT) can profoundly affect one or more of these activities, sometimes simply improving effectiveness, sometimes by fundamentally changing the activity, and sometimes by altering the relationship between activities. In addition, the actions of one firm can have knock-on effects for the value chains of customers and suppliers.

The standard value chain of the firm is illustrated in Figure 2.1. Within the value chain a number of examples of wireless applications are given. Let us examine each activity in turn.

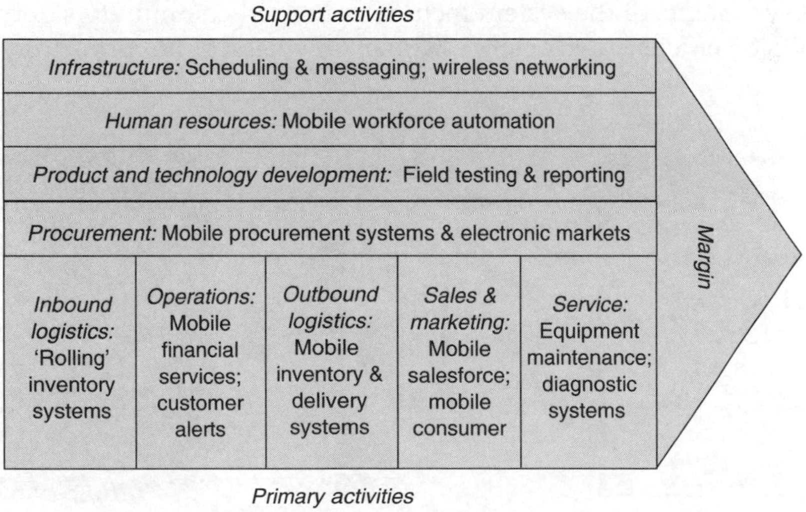

Support activities

Infrastructure: Scheduling & messaging; wireless networking					
Human resources: Mobile workforce automation					
Product and technology development: Field testing & reporting					
Procurement: Mobile procurement systems & electronic markets					*Margin*
Inbound logistics: 'Rolling' inventory systems	Operations: Mobile financial services; customer alerts	Outbound logistics: Mobile inventory & delivery systems	Sales & marketing: Mobile salesforce; mobile consumer	Service: Equipment maintenance; diagnostic systems	

Primary activities

Figure 2.1   Examples of mobile applications in the value chain of the business

Infrastructure

Wireless IT can provide significant business benefits for corporate infrastructure and a large number of corporate solutions have been developed to this end (Alanen and Autio, 2003; IBM, 2003; Microsoft, 2003). For example, advances in wireless messaging allow mobile workers to direct specific incoming messages to specific devices; such control helps mobile workers direct urgent e-mails to handheld devices and less urgent matters to secondary devices such as desktop PCs (Ferguson, 2000; Koehn, 2002). Other mobile office tools are also available – linking to fax, databases, schedules, and file transfer (Research in Motion, 2000a). Figure 2.2 demonstrates the linkage between fixed-wire and mobile information systems. In this scenario, an incident report entered into a system via PC, typically by the call centre, is automatically routed to the device of an appropriate engineer. The information on the incident and customer details appears on the PDA display – in this case a 'valve leak' that must be repaired as a priority.

On a much broader level, wireless networks and devices can help to strongly integrate remote, disparate, or roaming employees into the corporate infrastructure. For example, the Montgomery County Division of the Maryland National Capital Park Police have recently adopted Blackberry PDAs as a medium to integrate remote officers spanning 432 parks over 29,000 acres into the information network. Key benefits of the system include enhanced community safety and officer efficiency (Hamel, 2002). In businesses, important links to

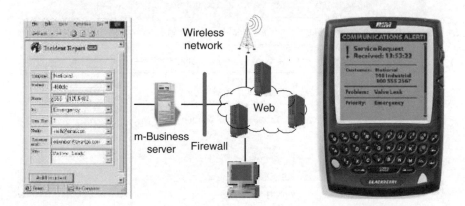

Figure 2.2 Messaging alert directed from a PC to a mobile device

company networks and systems such as Enterprise Resource Planning (ERP) can be facilitated (Bevis and Patterson, 2002; IBM Pervasive Computing, 2000c,d). Thinking more generally about the mobile workforce, employees are enabled to work in their virtual office at any time, in any place, and anywhere. Ovum predicts that the number of mobile-office users will grow to 137 million by 2004, up from 15 million in 2000 (Autio *et al.*, 2001).

One organization where mobile devices have been employed to enhance organizational infrastructure is the California Department of General Services – the general administrative office for California's state government. Here, more than 500 PDAs have been deployed to enable greater productivity away from the office (Microsoft, 2002a). In addition to general contacts and access to documents, a number of vertical applications have been provided for basic business operations. One such example is a system for assisting in the monthly automobile auctions for the Fleet Administration, where vehicles are sold at the end of their life cycle. Using a wireless local-area network (WLAN), based on the IEEE 802.11 standard, auction agents have detailed information at their fingertips in the auction lot, replacing large volumes of paperwork. The information is accessed via the vehicle ID number. Once the winning bid is taken, the information is sent wirelessly to a booth on the lot where all the pertinent information and forms are completed, including title transfer. The system has increased the number of vehicles that can be processed at auction, doubling the monthly sale of automobiles, reducing inventory, and increasing revenues (Microsoft, 2002a).

One frequently cited example of mobile networking is that of the businessperson in a car or taxi, or at the airport or rail station (Research in Motion, 2000b). Imagine the businessperson receives a call and is needed urgently in another city. Accessing a travel portal on a mobile device such as a PDA, the employee enters the preferred times of departure and return, requests a hotel near the visited office, and orders a conference room, catering facilities, and a presentation link back to the main office. Automatically, in the background, the system makes the required reservations using known preferences for the airline, in-flight meal, seating, rental-car, and so on (HP, 2000). In the situation where the businessperson is in a car, an in-vehicle network, for example, based on a standard for short-range connectivity such as Bluetooth, could allow connectivity of personal devices; using a mobile phone as the wide-area network connection, the employee could connect to the Internet or corporate intranet to receive information such as e-mail (Miller, 2000). Nevertheless, such ideas do beg the question of whether the businessperson can safely perform the main activity, namely driving. Advances in voice-recognition technology promise to help even further in this regard.

Human resources

Human resource management is not an area that springs immediately to mind when one thinks of wireless IT applications. Handheld training devices may be useful for remote or roaming workers. We could perhaps envision employees being electronically 'tagged' and monitored in a manner akin to some well-known science fiction novels, although providing interesting implications for civil libertarians. Nevertheless, some of these wireless ideas have been productively employed for organizational benefit and are described in other chapters of this book.

In the healthcare industry, an interesting workforce automation application is that of Rensimer's e/MD^2, shown in Figure 2.3 (Microsoft, 2002b). A major part of the effort to control medical costs is quantifying clinical work performed by physicians. Doctors typically use a complex system of service codes, based on the International Classification of Diseases, to be reimbursed for their services. Paper-based systems are slow and error-prone because of the numerous diagnosis/medical procedure codes that non-medical clerical workers must wade through to translate a doctor's notes into accurate billing and medical records. Rensimer's application allows doctors to enter the elements of their work using a stylus on a touch-screen Handheld PC to derive diagnosis codes for creating billing codes on

Figure 2.3 The e/MD^2 application (Microsoft, 2002b)

the spot. The application takes advantage of the familiar Windows interface to eliminate the learning curve for computer-wary physicians. The e/MD^2 application can eliminate coding paperwork by up to 90 per cent, saving thousands of dollars a year. Higher accuracy rates resulting from e/MD^2 mean doctors can provide more accurate billing records/claims to insurance companies, ensuring more complete and accurate patient medical charts or records (Microsoft, 2002b).

Product and technology development

Product and technology development is one of the activities in the value chain where the impact of wireless IT is more embryonic. There are currently few areas where workable solutions are being used. However, one area where wireless IT is likely to have an important role is in field testing and reporting (Research in Motion, 2000a). As the cost of wireless transceivers falls and the performance of said devices increases, impetus is provided for the inclusion of these chips in products such as cars, refrigerators, washing machines, vacuum cleaners, industrial equipment, and many other devices and appliances (US Internet Council and ITTA, 2000). Such devices will be able to store and report information on the performance of products, providing an important source for future product and technology development and improvement. Several forward-thinking companies such as Ariston, Dyson, Philips, and Ford have already expressed an interest in the inclusion of such wireless devices in future product offerings.

Procurement

Although purchasing products or services over mobile devices is expected to be an important part of e-commerce in the B2C market, it is difficult to imagine how procurement might occur using mobile IT in the business-to-business (B2B) domain – or at least it is in the developed world. In most developed economies desktop computers have become firmly rooted in the B2B procurement process, and it is not easy to see where mobile IT might fit in – except perhaps for exceptional roaming employees who are involved in procurement.

Nonetheless, in the developing world, the situation is very different. China provides a good example – although eager to become a player in the global B2B market, the country suffers from a dearth of PCs, and this is seen as a huge potential stumbling block (*E-commerce Times*, 2000). Moreover, only 10 per cent of medium-sized businesses are online. However, interestingly, the saturation of mobile technology for business is very high and most Chinese businesspeople carry a mobile phone or pager. This provides the potential platform for unprecedented B2B mobile commerce in China and procurement is likely to play a key role (Brown-Kenyon and Perkins, 2000). For example, a recently announced system involving Motorola, the China Wireless Information Network, and MeetChina.com is aimed at creating a huge platform for B2B procurement. The system will send B2B purchase inquiries via mobile devices to 2 million Chinese manufacturers and traders (*E-commerce Times*, 2000). Initially, the service provides industry information on five different channels, but its founders hope that it will evolve into a real-time electronic marketplace.

Inbound logistics

Using wireless IT, inbound inputs to the firm can be accurately monitored. Wireless transceivers let portable terminals – such as PDAs – communicate with a central database. Terminals can log in shipments of materials from vendors and track those materials in inventory, as they are needed (Zeus Wireless, 2002a). In this situation it is even conceivable to know all inventory in transit – or 'rolling' inventory – allowing an efficient method of selecting a source of components based on their known location. By knowing the location of 'rolling' inventory, times between transaction, manufacture, and delivery can be further reduced (Varshney, 2000).

A good case study where wireless IT has been used to facilitate ordering and delivery is that of the Johnson & Johnson subsidiary Ethicon – one of the world's largest manufacturers and suppliers of sutures and wound closure products (Hurwitz Group, 2001). Ethicon has recently provided a new automated stock and inventory system that ensures that UK hospitals' health supplies remain fully stocked. The solution, E-sy scan, allows the whole supply chain – from the hospital supply cupboard to Ethicon supply depot – to be seamlessly activated to ensure that critical surgical supplies arrive in the right place at the right time (Hurwitz Group, 2001). Handheld devices – Symbol's SPT 1500 equipped with a barcode reader – are used by the customer and link electronically to a JD Edwards ERP system.

Operations

The impact of wireless IT on the operations component of the value chain is likely to be enormous; there are vast arrays of applications under development and new ideas constantly appear on the horizon. Examples include:

Meter reading. In the utilities industry (e.g. gas, electricity, and water), meter reading tends to be a drain on time and resources for field staff. London Electricity pays £26/year per customer to its outsourced meter reading company. Wireless reading systems take a giant leap over this problem. For example, electricity meters installed at customers' homes could be read remotely by an operator working at his desk in the main office or even automatically (Research in Motion, 2000b). There are 600 million gas, water, and electricity meters in the United States and Europe; by 2003, one in four of these meters may incorporate a wireless device for automatic reading (Autio *et al.*, 2001).

Customer alerts. Advanced messaging systems, such as EnvoyWorldWide, MessageMachines, and Unimobile can help to alert customers to certain important information. For example, Network Associates – a US computer security firm – use a messaging system to efficiently alert its customers of a computer virus threat by determining ahead of time who should be notified by pager, who by cell phone, and who by e-mail. As long as the digital addresses are determined in advance, the virus alert could be sent to numerous devices concurrently (Ferguson, 2000).

Alarm monitoring. In the public services, such as fire and police, automatic alarms can alert services to a call-out situation. For example, a break-in could trigger an intruder alarm, instantly informing the police and alerting a nearby officer.

Perimeter security. Wireless transceivers, communicating with a central PC or controller from various locations, allow security firms to reliably secure indoor and outdoor facilities (Zeus Wireless, 2002a).

Mobile credit authorization. In the taxi trade, for example, in-vehicle terminals can obtain credit card authorization over wireless networks (Research in Motion, 2000a).

Automated teller machines (ATMs). Similarly to vending machines, ATMs in remote locations can be connected to bank networks using wireless IT, allowing seamless transactions and reporting of cash inventories (Arthur D. Little, 2000).

Restaurant operations. Wireless technology significantly improves the efficiency of the ordering process for restaurants. Little Chef, one of the UK's major roadside restaurants, introduced a system in 2002. Waiters use handheld computers equipped with touch screens that enable them to take customer orders and send this information automatically to the point of sale and the kitchen. Key benefits include increased orders, staff efficiency, less wastage of food, and customer satisfaction from improved service (Microsoft, 2002g).

Financial services. Key operations of financial services can be facilitated over mobile devices. For example, companies can offer brokerage functions over mobile devices (IBM Pervasive Computing, 2000b); customers of brokers as diverse as Fraser Securities (Singapore), Fidelity (United States), and Fimatex (France) can now trade stocks using a mobile phone or PDA. A personalized portfolio can be offered along with quotes, buying and selling functions, stock history, market news, and so on. Similarly, banks can offer customers a variety of features over a mobile device. Most of the major banks in Hong Kong, for example, now offer basic mobile information services and some transaction functionality; American Express, Citibank, Dao Heng Bank, HSBC, and Hang Seng Bank offer account enquires, fund transfers, bill payment (including credit card), and in some cases securities trading and access to foreign exchange markets (Ghani, 2001; Maude *et al.*, 2000).

Outbound logistics

Similarly to inbound logistics, wireless IT – especially location technologies – can play an important part in outbound logistics. Intelligent transportation systems are being introduced around the world, and wireless computing plays a key part in almost every solution. Taxis are being equipped with automatic vehicle-location devices, allowing the dispatch system to automatically select the taxi closest to the pickup location. Similarly, fleet-management systems are helping freight companies to monitor the status of deliveries and other outbound logistics activities. FedEx and United Parcels Service both use such systems. For example, IBM's route sales solutions allow adjusting and printing invoices from a vehicle, driver access to customer details, and the tracking of outbound goods (a handheld system can keep track of deliveries, pickups, and returns for end-of-route reconciliation) (Farrell, 2002). Conceivably, automatic inventory scanning and calculation within a vehicle is possible, computerising the process even further.

During 2002, Tesco.com, the e-commerce arm of Tesco, the UK's largest food retailer and supermarket chain, introduced a wireless system for extending their automation structure to their delivery vans. While Tesco already had experience with wireless-connected tablet-style PCs in their stores, they wanted to place more rugged, less-expensive devices in their delivery vehicles. Tesco built a prototype solution in-house that downloads the order list for a vehicle in the store over a WLAN (IEEE 802.11b) network, and assists the driver on the road using a Pocket PC (Microsoft, 2002c; see Figure 2.4). In this way, much of the delivery paperwork was replaced (see Figure 2.4a), and the driver is presented with a map plotting the course from the store to the customer's home (see Figure 2.4b). In addition, the system lists product substitutions for out-of-stock items, customer acceptance or rejection of these items, the customer signature, and routing and parking information for each customer in an effort to increase accuracy of delivery times.

One major inventory application for wireless is in vending (AM Vending Group, 2002; DMS Tech, 2002; Zeus Wireless, 2002a,b). Typically, a vending machine in a remote location could automatically signal to the corporate computer when it requires restocking (AM Vending Group, 2002), and this information could be immediately used in the supply chain to source vending products from

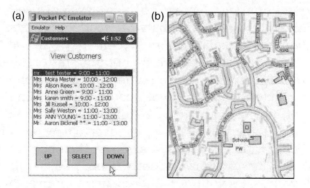

Figure 2.4 Tesco.com's wireless delivery application (Microsoft, 2002c): (a) schedule of customer deliveries; (b) delivery map

suppliers. Restocking of vending machines, for example, in a large office building, can be managed quickly and efficiently, maximizing cash collection. In addition, online communications with vending machines allows activities in other parts of the value chain including cash inventory, systems status, sales information, changing product prices, and updating software (Zeus Wireless, 2002b).

Sales and marketing

Applications of direct marketing and selling over mobile devices, particularly in B2C markets, are well known in the wireless literature, and numerous companies now have m-commerce capability, including Amazon, BarPoint, ESPN, E-Trade, Fidelity Investments, and Ticketmaster (Bergeron, 2001). Numerous other creative applications are on the horizon, including Coca-Cola's C-Mode – a next-generation vending machine with a barcode reader and video screen. When linked with I-mode, the machine becomes not only a drinks dispenser, but a multimedia terminal, capable of ticketing and many other m-commerce functions (MacDonald, 2003).

Retailing is an area that has had mixed fortunes in the e-commerce sector, with successes and failures alike. Nevertheless, it is an area where some mobile commerce applications are being developed and piloted. One recent application is that of online grocery shopping with Safeway using a PDA, IBM's Easi-Order system, and Intelligent Miner software (IBM Pervasive Computing, 2000a). Customers use Easi-Order to place grocery orders and interact with Safeway at virtually any time and from any place using a wireless modem connection (see Figure 2.5). The system uses a specially adapted PalmPilot (fitted with barcode readers for possible use in store or at home). Using data mining technology, Safeway's new system analyses a customer's 3-month shopping history and provides a suggested shopping list. It also recommends new products based upon prior purchases and those of customers with similar profiles. It knows not to offer meat to vegetarians, or dog food to people without pets, but it will offer baby products to new parents. Once an order has been made it is collated and can then be picked up from the local store. The system is currently being piloted and reportedly increases loyalty, sales volumes, and creates a more enjoyable shopping experience (Jarrett, 2002). Another supermarket application involves the use of wireless transceivers to connect point-of-sale

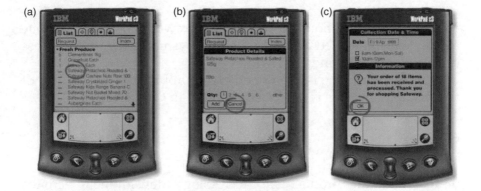

Figure 2.5 Safeway's Easi-Order application on the Palm PDA: (a) the product list; (b) product details; (c) sending an order

(POS) terminals to a central PC or controller without a wire. This lets merchants easily add or move POS terminals or printers (Zeus Wireless, 2002a).

National Express in the United Kingdom recently introduced a mobile solution to manage the delivery of both tickets and timetable information to train passengers, issuing 1500 PDA units to on-train staff. The data on the handheld device are updated overnight, and a portable printer/credit card reader allows for the easy management of ticket issue and payment processing. The system provides better customer service, reduces fraud, and improves the accuracy of ticket pricing (Microsoft, 2002f).

Besides consumer sales, mobile IT has other distinct applications for the business. In many industries, the salesforce is becoming increasingly mobile and teleworking is a very real part of sales activity. Wireless computing provides an important progression of the current technologies in this area and allows strong integration of a remote salesforce into ERP and other key systems (Jarrett, 2002).

Cybex International – a high-end sports equipment maker – provides a good example of sales and marketing applications in the value chain. Cybex had a common problem; its large salesforce visited customer locations but had very little idea of the number of products – in this case sports and fitness machines – it had available to sell. This information was stored back in the company's PeopleSoft ERP system (Nelson, 2000). By implementing wireless technology to give its salespeople quick and ready access to the information they needed on-site, these individuals could work far more efficiently and effectively and improve customer service. The chosen wireless solutions package for Cybex was PeopleSoft's Mobile eStore, which allowed access to the ERP system using any wireless device that could handle HTML.

Wireless marketing is another emerging application in this area of the value chain. This is described in detail in Chapter 8.

Service

Similarly to the product and technology development activity, devices can be embedded in products to bring benefits to the service activity. For example, an elevator under warranty may indicate to the manufacturer that it requires a free service. An oven might call General Electric when a heating element is about to expire (Shankland, 2000).

For physical products, the after-sales service element of the value chain often involves the use of field technicians. One example is Mesa Energy Systems – a full-service heating, ventilation, and air conditioning (HVAC) service and repair company. Mesa Energy is in a highly competitive business where communications traditionally have included relatively inefficient radio dispatch and paper processes. Mesa now uses a computer-aided, automated wireless-dispatching service to allocate jobs to the nearest trained field technician and to update the customer's record after the job is done. Using PDAs, technicians can rapidly access information needed for doing their jobs, and data sent via wireless transmissions helps communicate more effectively with customers and improve the progress of jobs (Microsoft, 2002d). For example, if a field service engineer is sent to the customer to investigate a problem, then using a wireless computer he can immediately determine the availability of a part needed for repair while the customer waits. Overall, the system has created an estimated 15 per cent improvement in the efficiency of a technician's workday and a 6-hours per week reduction in the time dispatchers need to communicate with technicians. Full information on every job gives technicians the ability to better manage their time, providing faster responses to customer requests, detailed service reports within minutes of job completion, and accelerated billing cycles (Microsoft, 2002d).

Another example is that of Xerox Global Services. Xerox use a central Asset Management system to track all network printers and copiers on behalf of their customers. Traditionally, the field technicians

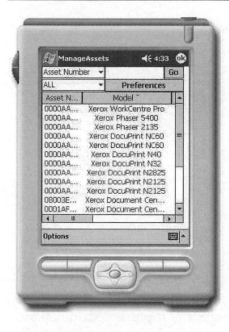

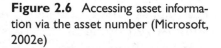

Figure 2.6 Accessing asset information via the asset number (Microsoft, 2002e)

carried information printouts with them to customer sites, bringing back handwritten meter readings and entering the readings into the computer manually. This is a labour-intensive, error-prone process. Using a Pocket PC, technicians are now able to access asset information and make necessary updates without having to wait until the end of the day or writing things down on paper (see Figure 2.6). The application uses a barcode reader to capture asset numbers, and uses 802.11 wireless networking when it is available and a cradle when it is not. Technicians can also collect meter readings for non-networked copiers using the Pocket PC, and view the service history for a particular asset. At the same time, technicians can use the Pocket PC to receive e-mail alert notifications from networked devices that have gone down (Microsoft, 2002e).

Assessing the impact of wireless applications on the business

There are a variety of key benefits that may be identified as business activities become 'unwired'. Many of these have been demonstrated in the assortment of wireless applications surveyed above. This

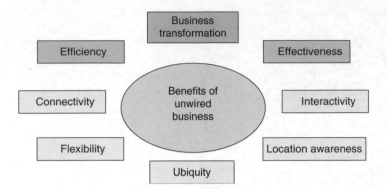

Figure 2.7 Key benefits of unwired business

section assesses a number of these organizational benefits. To this end, eight of the core benefits of wireless business applications are summarized in Figure 2.7. These items are not mutually exclusive. The three items at the top of Figure 2.7, represented in a darker shade, are generic business benefits from IT applications (Farbey *et al.*, 1992). At the highest level is *business transformation*, typically by process integration, process redesign, or adjustment of business scope. Business transformation is based on gains in *efficiency* and *effectiveness*, both important business benefits in themselves. The five items at the bottom of Figure 2.7 are specific benefits associated with wireless information technologies feeding into efficiency and effectiveness gains, as demonstrated by the examples in the value chain. Here, we find *connectivity, flexibility, ubiquity, convenience,* and *interactivity*. Let us explore these benefits in more detail.

Wireless IT can profoundly affect many business activities, sometimes simply improving efficiency, sometimes by fundamentally changing a particular activity, and sometimes by transforming the relationship between activities. In addition, the actions of one firm can have knock-on effects for the value chains of customers, suppliers, and other entities. In the above examples, the benefits that accrue in implementing and using wireless IT occur both in terms of *efficiency*-based cost savings via automation, for example, meter reading, and *effectiveness* via process transformation, for example, logistics. Both present attractive incentives to adopt wireless business services. However, it is also clear that, as in traditional use of IT, *transformation* benefits will prove the most elusive, requiring a significant amount of thinking 'outside of the box'. This can be illustrated

Table 2.1 Applications of wireless IT and business transformation

Level of transformation	Applications of wireless IT
Localized exploitation	Automation of specific business tasks. For example, automated meter reading to save costs
Internal integration	Networking and sharing information. For example, the mobile office
Business process redesign	Transforming sets of business processes. For example, using wireless IT in logistics to allow 'rolling inventory', thereby cutting times between orders, manufacture, and delivery
Business network redesign	Transforming relationships with other entities. For example, wireless marketing that targets customers via profiles and context-dependence (e.g. time or place)
Business scope redefinition	Creation of new revenue streams. For example, mobile operators leveraging wireless technologies to become banks (e.g. storing cash via mobile devices or networks)

using Venkatraman's (1994) model of IT-enabled transformation. The model suggests that there are five key layers of IT-enabled transformation, ranging from localized exploitation to business scope redefinition. Table 2.1 demonstrates some of the possible applications of wireless IT in each of the levels. The more the business uses IT to transform, the more benefits can be achieved. As such, companies that are willing to take risks in adopting wireless technology and using it creatively have tremendous possibility for achieving strategic advantage in the marketplace. However, the novelty, risk, complexity, and cost of the largest transformations will certainly prove elusive to many firms.

Turning to the specific benefits of wireless technologies, we find a range of items. Typically, the technologies offer a high degree of *flexibility* in the way that they are used. The small size of wireless chips means that they can be embedded in all manner of devices to provide a new wave of adaptability in the way that people use them. Associated with this, the designs of products and solutions become more flexible, without the need for proprietary data cables, such as in security or previously wired telemetry applications. On another level, organizations, such as offices or supermarkets, can easily rearrange IT equipment without significant cabling issues.

One of the key watchwords associated with wireless communications is that of mobility; wireless can be used 'on the move'. As long as the network covers the wireless transceiver, there is a *ubiquity* in data communication, indoors and outdoors. Whatever task or process is involved, there is the potential, at least, for mobile data communications. With the growth of mobile telecommunications in consumer markets, ubiquity is growing fast. Another aspect of ubiquity is the availability of generic standards for communication, such as Bluetooth and WLAN. Ubiquity is a key aim of players in the Bluetooth Standards Industry Group (SIG) and other standards groups. The diversity of product offerings from companies like those in the Bluetooth SIG (e.g. mobile phones, PDAs, computers, computer hardware, and software) creates a strong platform for market penetration.

One of the main espoused benefits of wireless solutions is simplicity. In the current 'wired world', connectivity of electronic equipment is dictated by cumbersome and complex wiring solutions. Removing the wires makes connection simpler; for example, automatic *connectivity* can be granted to any device in a network (such as a PDA or phone), and a large number and variety of devices are supported. In theory, the user no longer has to think about the problems of establishing device networking, only its benefits; any device embedded with a suitable short-range wireless unit can link into a network and begin sharing data. Even traditionally isolated devices, such as vending machines, can be connected effortlessly to provide additional services (such as information on current performance or maintenance requirements).

The potential for complex *interactivity* and information sharing between devices is enormous. Most of the examples described above indicate the need for interactivity in data communications, at varying levels of complexity, from simple telemetry to advanced access to corporate systems. For the user, services can be customized to specific needs. Further, devices will become more interactive and software technologies are being developed to allow a fuller understanding of capabilities, e.g., Jini (Shankland, 2000).

Advances in network location technologies and the derestriction of the global position system (GPS) have opened the door to a plethora of applications integrating *location-awareness*. Such standards allow location positioning up to 10 m. Standards bodies such as the Location Interoperability Forum have been established to enable the rapid commercialization of compatible location products. Such products can contribute to advanced offerings in logistics and many other areas.

Summary and conclusions

Wireless technologies provide a powerful platform for the development of a vast array of strategic applications in the value chain, a number of which may prove valuable in achieving competitive and other benefits. No part of the value chain remains untouched in our analysis; every primary or support activity in Porter's value chain has one or more clear applications of wireless IT. The key watchwords for all of these technologies and applications are portability and networking – significant computing trends since the 1990s. By leveraging the benefits of wireless technology, such trends can be taken to new heights.

Of the recent technological advances, embedded wireless systems and LBS appear to be important in many of the applications; from monitoring 'rolling' inventory in inbound logistics through to diagnosing product faults in the service activity, or from wireless office networks as part of infrastructure to automating a remote workforce in the human resource management activity. Seemingly, where there is information to be exchanged, there is a role for wireless IT. Industry sources forecast that by 2003, the number of wireless computing devices will exceed the population of our planet – estimated at 6 billion. This figure is predicted to consist of a mixture of: around 300 million PDAs; approximately 2 billion consumer electronic devices such as wireless phones, pagers, and set top boxes; and 5 billion additional everyday devices like vending machines, refrigerators, and washing machines embedded with chips connected to the Internet (IBM Pervasive Computing, 2000d).

Nonetheless, it is a mistake to imagine that all of the wireless applications illustrated will succeed. Time will tell which wireless technologies or applications become pervasive and dominant. Many standards have yet to become fully developed or accepted and some of the more powerful technologies – such as fast, 3G networks – have yet to become offered fully. However, it is clear that wireless IT has an important part to play in the IT strategies of many organizations in the future; a large number of the applications presented in this chapter are either being successfully used or piloted in organizations today. With further technological advances, we are likely to encounter even more creative and interesting applications in the value chain.

Acknowledgements

An earlier and shorter version of this paper appeared as: Barnes, S.J. (2002). Unwired business: wireless applications in the firm's value chain. *eBusiness Strategy Management*, **4**, 27–37.

References

Alanen, J. and Autio, E. (2003). Mobile business services: a strategic perspective. In B.E. Mennecke and T.J. Strader, eds, *Mobile Commerce: Technology, Theory and Applications*. Hershey: Idea Group Publishing, pp. 162–184.

AM Vending Group (2002). Wireless data. http://www.amonline.com/ tools_of_trade/wire_data/index.shtml, accessed 20 December 2002.

Arthur D. Little (2000). Serving the mobile customer. http:// www.arthurdlittle.com/ebusiness/ebusiness.html, accessed 15 January 2000.

Autio, E., Hacke, M., and Jutila, V. (2001). Profit in wireless B2B. *McKinsey Quarterly*, No. 1, 20–22.

Bergeron, B. (2001). *The Wireless Web: How to Develop and Execute a Winning Wireless Strategy*. New York: McGraw-Hill.

Bevis, D. and Patterson, L. (2002). *Extending Enterprise Applications to Mobile Users*. Somers NY: IBM Corporation.

Brown-Kenyon, P. and Perkins, T. (2000). China: data in the air. *McKinsey Quarterly*, No. 4, 15–18.

DMS Tech (2002). Vending machine remote control. http:// www.dms-tech.com/en/iov/index.shtml, accessed 20 December 2002.

E-commerce Times (2000). China leads the way on mobile e-commerce.http://www.nua.ie/surveys/index.cgi?f=VS&art_id= 905355548&rel=true, accessed 15 January 2000.

Farbey, B., Targett, D., and Land, F. (1992). Evaluating investments in IT. *Journal of Information Technology*, **7**, 109–122.

Farrell, A. (2002). Mobile computing in the route sales marketplace. http://www-3.ibm.com/software/pervasive/tech/whitepapers/ mobile_computing.shtml, accessed 23 June 2002.

Ferguson, K. (2000). Messaging service focus on mobile workers. http://www.forbes.com/, accessed 6 September 2000.

Ghani, R. (2002). *The Future of Wireless Banking*. Somers, NY: IBM Global Services.

Hamel, D. (2002). Opportunities and challenges: building the wireless enterprise. http://www.discoverylogic.com/, accessed 1 September 2002.

HP (2000). Communications industry overview. http://www.hp.com/communications/, accessed 27 May 2000.

Hurwitz Group (2001). *Johnson & Johnson's Ethicon Limited Rolls Mobile Service into the Operating Room*. Framingham, MA: Hurwitz Group Inc.

IBM Pervasive Computing (2000a). IBM and Safeway create enjoyable grocery shopping experience. http://www-3.ibm.com/pvc/, accessed 28 November 2000.

IBM Pervasive Computing (2000b). Mobile solutions for e-business: financial services. http://www-3.ibm.com/software/pervasive/tech/whitepapers/mobile_finance.shtml, accessed 20 December 2002.

IBM Pervasive Computing (2000c). CRM/ERP/SCM. http://www-3.ibm.com/pvc/mobile_internet/crm_erp_scm.shtml, accessed 28 November 2000.

IBM Pervasive Computing (2000d). Extending SAP systems to pervasive computing devices. http://www-3.ibm.com/software/pervasive/tech/whitepapers/sap.shtml, accessed 20 December 2002.

IBM (2003). Pervasive computing. http://www-3.ibm.com/software/pervasive/, accessed 4 February 2003.

Jarrett, R. (2002). *Wireless Projects – How to Obtain a Return on Investment*. Basingstoke: IBM UK.

Koehn, D. (2002). Dynamic mobility: building the wireless Web. http://www.avantgo.com/, accessed 1 September 2002.

MacDonald, D.J. (2003). NTT DoCoMo's I-mode: developing win–win relationships for mobile commerce. In B.E. Mennecke and T.J. Strader, eds, *Mobile Commerce: Technology, Theory and Applications*. Hershey: Idea Group Publishing, pp. 1–25.

Maude, D., Raghunath, R., Sahay, A., and Sands, P. (2000). Banking on the device. *McKinsey Quarterly*, No. 3, 87–97.

Microsoft (2002a). California Department of General Services. http://www.microsoft.com/mobile/enterprise/casestudies/CaseStudy.asp?CaseStudyID=13423, accessed 20 December 2002.

Microsoft (2002b). Rensimer Enterprises, Ltd. http://www.microsoft.com/mobile/enterprise/casestudies/CaseStudy.asp?CaseStudyID=13362, accessed 20 December 2002.

Microsoft (2002c). Tesco.com. http://www.microsoft.com/mobile/
    enterprise/casestudies/CaseStudy.asp?CaseStudyID=13394,
    accessed 20 December 2002.

Microsoft (2002d). Mesa Energy Systems. http:// www.microsoft. com/
    mobile/enterprise/casestudies/CaseStudy.asp? CaseStudyID=13365,
    accessed 20 December 2002.

Microsoft (2002e). Xerox Corporation. http://www.microsoft.com/
    mobile/enterprise/casestudies/CaseStudy.asp?CaseStudyID=
    13400, accessed 20 December 2002.

Microsoft (2002f). National Express. http://www.microsoft.com/
    mobile/enterprise/casestudies/CaseStudy.asp?CaseStudyID=
    13428, accessed 20 December 2002.

Microsoft (2002g). Little Chef. http://www.microsoft.com/
    mobile/enterprise/casestudies/CaseStudy.asp?CaseStudyID=
    13352, accessed 20 December 2002.

Microsoft (2003). Mobile workplace. http://www.microsoft.com/
    mobile/enterprise/MobileWorkplace/default.asp, accessed 31
    January 2003.

Miller, B. (2000). Bluetooth applications in pervasive computing.
    http://www-3.ibm.com/pvc/tech/bluetoothpvc.shtml, accessed
    15 February 2000.

Nelson, M. (2000). Wireless ERP access empowers salespeople.
    http://www.planetit.com/techcenters/docs/enterprise_apps/
    news/PIT20000822S0014, accessed 21 August 2000.

Porter, M. (1980). *Competitive Strategy*. New York: Free Press.

Porter, M. and Millar, V. (1985). How information gives you compet-
    itive advantage. *Harvard Business Review*, **63**, 149–160.

Research in Motion (2000a). Building a business case for wireless.
    http://www.rim.net/, accessed 15 November 2000.

Research in Motion (2000b). The wireless workforce. http://
    www.rim.net/, accessed 15 November 2000.

Shankland, S. (2000). Jini's bottleneck. http://news.cnet.com/
    news/0-1003-201-1559726.html, accessed 15 March 2000.

US Internet Council and International Technology and
    Trade Associates (ITTA). *State of the Internet 2000*. Washington DC:
    ITTA Inc.

Varshney, U. (2000). Recent advances in wireless networking. *IEEE
    Computer*, **33**, 100–103.

Venkatraman, N. (1994). IT-enabled business transformation: from
    automation to business scope redefinition. *Sloan Management
    Review*, December, 73–87.

Wrolstad, J. (2002). New wireless tech pushes mobile sales apps. http://wirelessnewsfactor.com/perl/story/19888.html, accessed 6 November 2002.

Zeus Wireless (2002a). *Wireless Data Telemetry*. Maryland: Zeus Wireless Inc.

Zeus Wireless (2002b). Vending. http://www.zeuswireless.com/applications/vending.php, accessed 6 October 2002.

Strategic wireless technologies

Chapter 3

Experiences with Japan's iMode service

Co-author: Sid L. Huff

Introduction

One country has rapidly adopted the wireless Internet like no other – Japan. In doing so, rightly or wrongly, Japan has become an exemplar market of the wireless Internet. While the United States has grappled with its fragmented set of mobile network and service standards and Europe has experienced slow adoption of the wireless Internet based on WAP, the diffusion of wireless data services in Japan has been phenomenal. As of 30 September 2002, Japan had just over 59 million users of the wireless Internet (Mobile Media Japan, 2002), from a cellular phone subscriber base of nearly 72 million. Internationally, Japan has the highest rate of Internet capability in mobile phones (82 per cent), followed by South Korea (59.1 per cent), and Finland (16.5 per cent) (Nua, 2002). The United States is ranked as sixth with 7.9 per cent. In Japan, the leading telecommunications provider, NTT DoCoMo, has approximately 59 per cent of all wireless Internet subscribers, using its popular iMode service.

WAP was hailed by some as the de facto global standard for wireless information and telephony services on digital mobile phones and other wireless terminals (AU System, 1999); however, clearly it is iMode that takes the lead in terms of the number of wireless

Internet users. The iMode platform, based on compact Hypertext Markup Language (cHTML), provides a compelling alternative to WAP. Building on its early success in Japan, iMode has begun a strategy of market entry into the United States, Europe, and parts of Asia through a set of key partnerships (*Business Week*, 2001; PMN, 2002; Tech Web, 2001; Unstrung, 2002; van Meighem, 2002). The results of these strategic partnerships are being closely monitored by the telecommunications industry.

This chapter examines these issues in detail. To begin, the next section explores the nature of the iMode phenomenon, examining its growth, and positioning iMode within the range of wireless services and technologies available in Japan. The chapter continues by analysing the reasons for the success of iMode in Japan, using the theory of technology acceptance. The chapter then explores the planned advancement and growth of iMode, including transference issues as iMode moves to new markets, and the introduction of more sophisticated network and application environment technologies. Finally, the chapter concludes with a summary and considerations regarding the future of iMode.

The nature of iMode

NTT DoCoMo, Japan's leading cellular phone operator, launched the iMode service in February 1999. The 'i' of iMode stands for Internet, information, interaction, and I, myself. iMode is a mobile phone service that offers continuous, 'always-on' Internet access based on packet-switching technology. The packet-switched network used is Personal Digital Cellular – Packet (PDC-P). Using iMode, data can be immediately sent and received by networked handsets – a radical departure from traditional circuit-switched network standards used in other parts of the world, which require the user to 'dial up' and 'log-on'. The speed of the standard service varies between 9.6 and 28.8 Kbit/s. Because iMode is always on, billing for usage is based on the amount of data sent and received, not time online, paying per packet (128 bytes) of information.

iMode-enabled Web sites utilize cHTML, a subset of HTML 4.0. The language is designed with the restrictions of the wireless infrastructure in mind, such as the limited bandwidth and high latencies of the networks, and small screens and limited functionality of the devices. By removing certain features of conventional HTML, such as tables and frames, the speed of content delivery can be substantially increased (albeit display sophistication reduced). The cHTML language also has additional features, such as a 'phone to' tag to directly dial a phone number from a Web page. The decision to select HTML as a base language has provided several benefits over rivals such as WML. Since HTML is the most widely used language, it has enabled companies to focus less on technological adaptation and more on the creative side of content development (MacDonald, 2003). This, coupled with the use of HTTP, has enabled content providers to interact with existing systems in a far smoother way. Arguably, the presentation of content is also more attractive, with a variety of input methods (e.g. radio buttons, drop down menus, and check boxes), and the language is more forgiving for the programmer (MacDonald, 2003).

Provision of Internet content to mobile users occurs via a specially enabled iMode phone and browser application. An iMode-enabled phone typically weighs approximately 90 g (3.6 ounces), has a comparatively large liquid crystal display (LCD), and a four-point command navigation button that allows the user to manipulate a pointer on the display. A typical display size is 120×160 screen dots or 10×10

(a) (b) (c)

Figure 3.1 Examples of iMode-enabled cellular phones (http://nooper.co.jp/showcase/): (a) Toshiba T210lv 3G; (b) NEC N2001 3G; (c) Panasonic 210lv 3G

characters. The user connects to the iMode service by pressing a single button. Examples of iMode phones are shown in Figure 3.1. Some recent phones also include a built-in camera, movie mail, and the ability to play MP3 files.

The client application used for delivery of information is a microbrowser, a simplified, compact version of a traditional Web browser such as Internet Explorer or Netscape. One popular browser, Compact NetFront, developed by the Japanese company Access, is currently used in 75 per cent of all iMode-enabled devices (Eurotechnology, 2002). The browser has two bars with icons at the top and bottom of the LCD screen. These can be customized to allow access to various services and menus. In the middle of the screen is the main display area that provides text and graphics (see Figures 3.2 and 3.3).

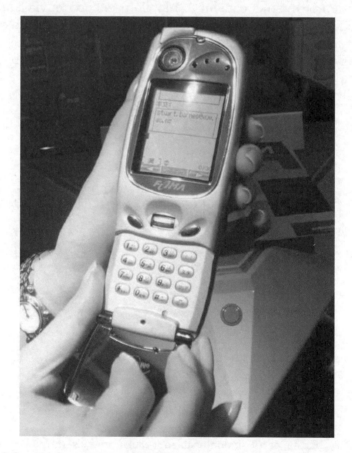

Figure 3.2   Sending an e-mail on iMode

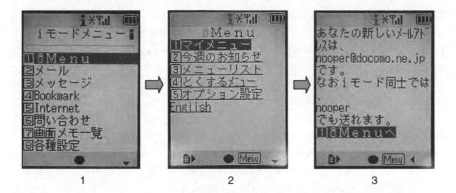

Figure 3.3 Example of the iMode micro-browser (http://nooper.co.jp/showcase/)

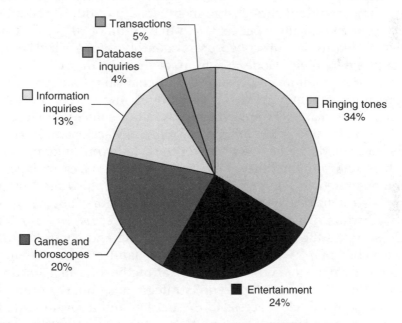

Figure 3.4 Key areas of usage for iMode services (TCAJ, 2002)

Typical services include e-mail, games, weather forecasts, sports news, restaurant information and coupons, transportation schedules, banking, data searches, and information updates. The typical user receives 5.1 e-mail messages per day, sends 3.9 of them, and browses 8.9 Web pages per day (Negishi, 2002). The current pattern of usage, based on 2002 data, is shown in Figure 3.4. As we can see, downloading ringing tones is the most popular area of usage, with around one-third of all iMode traffic, as users download tones to personalize their individual phones. This is closely followed by entertainment, with a quarter of all traffic, and a closely related segment, games and horoscopes, with a fifth. Clearly, the overarching emphasis of services is on entertainment, fun, and personalization.

DoCoMo originally controlled the provision of content for iMode through a portal page. Currently, there are approximately 3000 sites offered in this way via the iMenu, provided by nearly 2000 companies. Each of the links has been approved by DoCoMo for uniqueness and usefulness. Additionally, there are in the order of 50,000 'unofficial' sites created by non-partners. In March 2001, DoCoMo opened up iMode site selections to allow access to many of those unofficial sites, thereby providing additional revenue generation for the company. Some premium sites charge members a monthly subscription fee for content (usually 9 per cent, to a maximum of 300 yen), which is then added to the subscribed users monthly phone bill. The user typically enters their iMode PIN number through the cell phone to register for a premium site. DoCoMo automatically transfers fees to the premium content provider's bank account, minus 9 per cent taken as a handling charge. The market for premium content on iMode is estimated to be at least $60 million per month (MacDonald, 2003).

The basic model of the ecosystem for the iMode consumer market – which is relevant for other wireless Internet services in Japan – is demonstrated in Figure 3.5. (Note: a more sophisticated model might also consider system and network vendors, but these are assumed to be driven via DoCoMo and thus *ceteris paribus*.) NTT DoCoMo takes the central role in coordinating and controlling the whole value map, working closely with handset manufacturers, content providers, and others, to create attractive offerings for users and what it refers to as a 'win–win' strategic position. For example, DoCoMo provides clear access to the market and a revenue model for content providers with rich and attractive content; at the same time, DoCoMo does not have to pay for content, as is required by operators in some markets. The provision of handsets is heavily subsidised by the operator, who also defines and specifies handset features to the manufacturer.

Based on US accounting standards, NTT DoCoMo had consolidated net sales of 2,384,254 million yen for the 6 months up to September 2002, up 1.9 per cent year-on-year (Kawasaki, 2002). Pre-tax profit rose 22.3 per cent to 627,967 yen. However, net profit

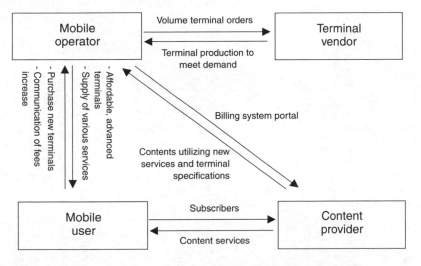

Figure 3.5   The mobile ecosystem for iMode (after Rauhala, 2002)

fell sharply to just 4.147 billion yen (a fall of 95.3 per cent), particularly due to overseas-related investment losses as iMode has struggled to enter into new markets (see below). For the full year to March 2003, NTT DoCoMo estimates that net sales will reach 4.676 trillion yen, up 0.4 per cent over the previous year. Pre-tax profit is expected to rise by 4.4 per cent year-on-year to 998 billion yen; net profit is forecast at 182 billion yen, rebounding from a net loss of 116.2 billion yen the previous financial year (Kawasaki, 2002).

Other wireless Internet services in Japan

Although iMode is by far the largest wireless Internet service in Japan, it is not the only service. There are two other major contenders for mobile data services in the country – EZWeb (owned by KDDI and delivered under the 'au' brand) and J-Sky (owned by J-Phone and in which the Vodafone Group has a majority shareholding) – each with its own distinct technological characteristics and market positioning (see Table 3.1).

The largest competitor to iMode is EZWeb, with 11.15 million users as of 30 September 2002 (Mobile Media Japan, 2002). EZWeb was originally based on a modified WAP standard, improved to include additional features such as colour. The original standard was based on Unwired Planet's Handheld Device Markup Language (HDML) and

Table 3.1 Comparison of major wireless Internet services in Japan

	iMode	EZWeb	J-Sky
Ownership	NTT DoCoMo	KDDI – 'au' brand	J-Phone/Vodafone
Markup language	Compact HTML (cHTML), a modified version of HTML 4.0	1. Handheld Device Markup Language (HDML) 2. Extensible Device Markup Language (xHTML), based on WAP 2.0 for 3G services	Mobile Markup Language (MML), a modified version of HTML 4.0
Main micro-browser	Compact NetFront	EZbrowser	Proprietary
Number of sites	3018 official sites, 53,736 unofficial sites (4/02)	N/A	N/A
Mobile Internet fee	1. Basic subscription of ¥300 monthly; ¥0.3 per packet (128 kB); up to ¥300 subscription for premium content on official sites 2. 3G service costs ¥100 more per month and ¥0.2 per packet	1. Basic service: ¥300 monthly and ¥0.27 per packet 2. 3G service: ¥600 monthly and ¥0.27 per packet	1. Basic service: ¥2 per 1 kB sent or received; no monthly fee 2. Packet service: Basic subscription of ¥300 monthly; ¥0.3 per packet
E-mail attachments	Graphics and audio	Graphics, audio, and video	Graphics and audio
Network	1. Circuit switched and packet switched network based on Personal Digital Cellular (PDC) 2. 3G network based on Wideband-Code Division Multiple Access (W-CDMA) introduced in October 2001	1. Packet switched CDMAOne network 2. 3G network based on CDMA 2000 1x introduced in April 2002	1. Circuit switched and packet switched 2. 3G network based on W-CDMA introduced in December 2002

Maximum speed	2G: 9.6 Kbit/s (circuit); 28.8 Kbit/s (packet) 3G: 384 Kbits/s downstream and 64 Kbit/s upstream (packet); 64 Kbit/s (circuit)	2G: 64 Kbit/s downstream and 14.4 Kbit/s upstream (packet) 3G: 144 Kbits/s downstream and 64 Kbit/s upstream (packet)	2G: 9.6 Kbit/s (circuit); 28.8 Kbit/s (packet) 3G: 384 Kbits/s downstream and 64 Kbit/s upstream (packet); 64 Kbit/s (circuit)
Colour support	Up to 262k colours	Up to 262k colours	Up to 262k colours
Recent features	Built-in video camera; movie playback; audio/music playback	Built-in video camera; global positioning; movie playback; audio/music playback	Built-in video camera; movie playback; audio/music playback
International roaming	Moderate. The World Walker service requires a special handset and contract with an international carrier. Has possible coverage in 110 countries	Limited. Global Passport service provides connectivity in specific locations such as: Hong Kong, major cities in Korea, the USA, New Zealand, and Australia, Beijing and Shanghai in China, and Vancouver and Alberta in Canada	High. The new 3G service provides a standard dual-mode handset option (W-CDMA/GSM) that will enable easy overseas roaming, particularly with Vodafone's expansive overseas networks

Handheld Device Transport Protocol (HDTP), which, respectively, describe the user interface and communication protocols for wireless devices. HDML is a forerunner to the Wireless Markup Language (WML), on which WAP is based. Since December 2001, KDDI has also been using WAP 2.0 on its CDMA2000 network.

Clearly, competition between EZWeb and iMode hinges in large measure on the WAP platform. Elsewhere, the WAP standard has met a rather apathetic response, and interestingly, although iMode has the lion's share of the market, Japan boasts a large number of WAP subscribers. Nevertheless, while WAP and iMode have different specifications at present, the WAP Forum have made the WAP 2.0 standard – which uses xHTML – iMode-compliant, resulting in an overarching application platform. Thus, for example, the SonyEricsson T68i, which uses WAP 2.0, can display both WAP and iMode sites. Until the standard becomes widespread, the platforms will remain in competition. EZWeb is increasingly gaining competitive advantage through its leading-edge phones, which are becoming more techno-logically sophisticated and with greater battery life. However, its inter-national roaming is limited, and the speed of the network, although enjoying better coverage, is slower than that of iMode.

A second competing technology for the wireless Internet in Japan is J-Sky, owned by Japan Telecom. J-Sky has a slightly smaller customer base than EZWeb, estimated at 11.08 million subscribers in September 2002 (Mobile Media Japan, 2002). J-Sky has been the least sophisti-cated platform of those available in Japan, and has only recently intro-duced a packet switching network. The service has traditionally been targeted at young female users, typically for messaging and e-mail communication. Since the network is not packet switched, J-Sky's pricing model has traditionally differed from other operators, being based on a fixed fee of 2 yen per request. Nevertheless, since Vodafone gained the controlling stake in J-Sky it has rapidly begun to gain ground on iMode and now offers similar services. Its new interna-tional roaming service – Vodafone Global Standard – promises to be the most expansive of all Japanese operators.

The growth of iMode in Japan

The growth and success of iMode has provided considerable food for thought for wireless Internet industry analysts. Some observers argue that the Japanese iMode example is unique and perhaps unlikely to be emulated elsewhere. However, there appear to be some key lessons that can be gleaned. iMode is a widely recognized brand, standing for concepts such as simplicity, functionality, and meeting consumer needs (Funk, 2000). This section explores the reasons for the success of iMode in Japan. Technology acceptance theory is applied to understand the underlying reasons for the rapid and successful customer adoption of iMode. Additional environmental factors that have contributed to iMode's adoption are also discussed.

Understanding iMode adoption using technology acceptance theory

Technology acceptance theory examines the factors that influence the adoption, and diffusion throughout a social system, of new technologies such as iMode. One theorist, Everett Rogers, has spent over 30 years studying the diffusion of innovations of all kinds, from the QWERTY keyboard to agricultural innovations in developing countries (Rogers, 1995). Rogers developed a set of key innovation characteristics that, as he and others have shown, serve to explain innovation diffusion outcomes very well:

> *Relative advantage*: the degree to which the innovation is perceived as being better than the practice it supersedes.
> *Compatability*: the extent to which adopting the innovation is compatible with what people do.
> *Complexity*: the degree to which an innovation is perceived as relatively difficult to understand and use.
> *Trialability*: the degree to which an innovation may be experimented with on a limited basis before making an adoption (or rejection) decision.
> *Observability*: the degree to which the results of an innovation are visible to others.

Other researchers have extended Rogers's work, suggesting additional factors to be considered along with Rogers's basic five, specifically image (the degree to which adoption and use of the innovation is perceived to enhance one's image or status) and trust (the extent to which the innovation adopter perceives the innovation provider to be trustworthy).

Starting at about the same time as Rogers, psychologists Martin Fishbein and Icek Ajzen conducted many years of research from which they eventually developed the Theory of Planned Behaviour (TPB) (Ajzen, 1991). This theory attempts to explain why people behave as they do in situations of 'reasoned action'. In the context of technology adoption, TPB posits that actual use of a technology is determined by the individual's *behavioural intention* to use the technology. Intention is, in turn, determined (in cases where use is voluntary) by the individual's *attitude* towards using the technology, and the *subjective norms* towards using the technology present in the individual's social milieu.

In recent years, Rogers's findings and those of TPB have been combined into a general theory of technology acceptance (Karahanna *et al.*, 1999; Venkatesh and Davis, 2000), illustrated in Figure 3.6.

This model can be applied to the case of iMode to help understand its adoption dynamics and success. First, the theory argues that an individual's attitude toward adopting iMode is determined by the key characteristics related to the technology: relative advantage, compatability, complexity, trialability, and observability, as well as trust and image. Let us examine each of these, in turn.

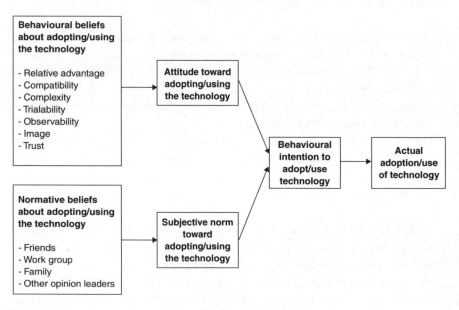

Figure 3.6   General model of technology acceptance

IMode's principal source of *relative advantage* stems from the fact that it provides an Internet access channel to many individuals for whom the Internet was effectively inaccessible previously. Before iMode, Internet penetration over traditional wired channels was constrained by the high online data charges set by DoCoMo's parent NTT, resulting in a high level of price-constrained, unmet Internet demand. Consumer enthusiasm for the iMode channel is evidenced in click-through rates for banner and e-mail advertisements, 3.6 and 24.3 per cent, respectively, compared with only 0.5 per cent for PC-based online banner ads (Nakada, 2001).

In terms of *complexity*, the iMode service has performed well due to its simple and intuitive interface of menus and the command navigation button. The iMode micro-browser is also a scaled-back version of traditional browsers, further simplifying the user's experience. In providing the iMode service, DoCoMo has emphasized simple content. DoCoMo's control of the portal page and approval control of third-party service providers' access also contributes to the management of iMode's complexity. In delivering iMode services, DoCoMo has tended to trade off richness for reach (Wurster and Evans, 2000), and this surfaces clearly in iMode usage patterns: simple entertainment represents more than 50 per cent of the traffic (e.g. games, ringing melodies, Bandai characters), followed by e-mail and news (Funk, 2000).

IMode, being based on the familiar mobile telephone handset, maintains a high degree of *compatibility* with the consumer's past experience. iMode is also highly compatible with the Japanese cultural values. Modern Japanese culture is well known for its enthusiasm for electronic devices, especially among Japanese youth. Indeed, Japan has been at the core of innovation and development for many electronic entertainment devices – mobile or otherwise – such as Sony's Playstation or Nintendo's Game Boy. Therefore, it is perhaps not surprising that entertainment applications have played an important part in the success of iMode.

The low initial cost of the service, based largely on a pay-as-you-use mechanism, allows for a high level of *trialability* before confirmation of adoption. Trialability is also enhanced through social networks. The low cost per message of an iMode telephone means that friends can easily share their devices for trialling.

The immediacy of iMode also creates a relatively high level of *observability*. iMode usage is highly interactive; iMode transactions are responded to nearly instantly. Observability is also enhanced through individuals witnessing others using iMode phones, which often occurs 'out in the open'. Observability is lessened somewhat due to the fact that some important aspects of the innovation, such as the network, are less visible, and that many of the applications on iMode are rather abstract – such as news retrieval. On the other hand, iMode usage often forms the basis for observable behaviour: for example, restaurant information and booking, or person-to-person e-mail communication.

The iMode service clearly engenders a high level of *trust* (and thus low perceived risk) due to its large, established brand and owner, NTT DoCoMo. As well as NTT DoCoMo being an established and respected voice provider, iMode has become a recognized brand for mobile data services. Consumer trust in iMode is further reinforced through the use of the 'nttdocomo.ne.jp' e-mail address.

Related to visibility, and similarly to other mobile phone markets, the *image* of iMode devices is very important. Indeed, as in the European and US markets, many users select handsets on the basis of enhancing the individual's social status or image (Matsunaga, 2001).

Second, an individual's adoption of iMode will be affected by *subjective norms* toward using the technology. These norms are determined by normative beliefs attributed to significant others (friends, work colleagues, family members, and the like) with respect to adopting or continuing to use the technology. The Japanese cultural tendency toward group conformity (Asai, 2001) provides a strong basis for these norms. Once a new technology reaches a critical mass of usage, these cultural norms tend to accelerate adoption and continued usage. In that regard, iMode adoption is strongly centred in the youth and young adult age segment; approximately 70 per cent of iMode subscribers are under 35, and the segment who run up the highest bills are under 25. Also, loyalty to the NTT DoCoMo brand, discussed earlier, comprises a subjective norm which strengthens iMode adoption and use.

Other factors affecting iMode adoption and use

Technology acceptance theory operates principally at the level of the individual. However, in the case of iMode, there are some key factors in the economic and technological environment that also play an important role.

Market situation: NTT DoCoMo was – and still is – market leader, with 60 per cent market share in the mobile phone market. The majority of

DoCoMo's shares are owned by NTT, and the Japanese Government owns the majority of NTT's shares. This provides a monopolistic situation where the Government has a controlling stake, although successive waves of deregulation by the Ministry of Posts and Telecommunications promise to promote socio-economic structural reform.

Vertical integration: DoCoMo has a strong position in the mobile value chain, being vertically integrated into chip, handset, and infrastructure research and development. This power has been leveraged substantially. For example, DoCoMo received preferential access to the lightest phones from four suppliers in exchange for preferential information about new phone standards. Subsequently, these phone suppliers used this information to obtain strong cooperation from parts suppliers to make better design tradeoffs. Similarly, the release of Fujitsu's 256-colour screen phones in December 1999 allowed DoCoMo to support its own version of the Internet with attractive phones at affordable prices.

Network investment: DoCoMo has been proactive in its investment in network infrastructure. Its implementation of a packet data overlay allowed for spectrum efficient, cost-effective, key-push services, approximately one year ahead of Japanese rivals. The huge success of this relatively low-speed network between 1999 and 2001 has challenged the myth that new mobile Internet services cannot grow and succeed without high-bandwidth network standards – a position taken by many European and US analysts.

Self-reinforcing services: recently, it has become evident that there is a 'virtuous circle' connecting DoCoMo's voice and data services. While the average revenue per subscriber is lower for iMode than voice (¥2150/month compared with ¥7800/month), iMode users tend to make more voice calls, driven by the information that they access through the service, such as the names of certain restaurants.

In summary, the theory of technology acceptance helps us see how features of iMode technology, factors underlying the surrounding behavioural norms, and market and industry features have driven the rapid rise of iMode in Japan. How effectively have these forces combined together? As its peak, around 55,000 new users subscribed to iMode, indicating high behavioural intentions to use the technology. Market research also indicates a continuing steady stream of subscriptions that have so far outstripped all predictions (Mobile Media Japan, 2002).

Experiences with iMode outside Japan

Building on its early success in Japan, as early as 2000 NTT DoCoMo planned to export its iMode model overseas, particularly to selected markets in the United States, Europe, and elsewhere in Asia, where on 20 June 2002, the company launched iMode in Taiwan, in association with KG Telecom. Through a strategy of partnering, NTT DoCoMo hoped to emulate its earlier achievements in Japan. However, the company has seen a considerable loss in overseas investments in 2002 in the wake of a downturn in the telecommunications market.

The company began offering a European version of iMode for the German and Dutch markets in early 2002 (followed soon after by Belgium), and has since begun to move into a number of other European markets. The service was launched in France by Bouygues Telecom in December 2002, in an attempt to boost subscriber numbers (van Miegham, 2002). The Spanish operator, Telefonica Moviles, plans to introduce a service to the Spanish market based on iMode in April 2003 (PMN, 2002). There are similar plans for the United Kingdom and Italy through Hutchinson 3G UK (Sugiyama, 2002). Applications include LBS such as 'smart' car parking and traffic navigation, e-mail, and interactive games. Key to the entry of iMode into the European market is the availability of suitable handsets that support both WAP and iMode.

So far, the response to the European iMode service has been somewhat apathetic. The service – offered in partnership with KPN Mobile in the Netherlands and E-Plus Mobilefunk GmbH in Germany – had planned to reach a subscriber base of 1 million by 2003. As of December 2002, the total number of users stood at just 143,000, with an average monthly net addition of less than 15,000 subscribers (Unstrung, 2002). One key problem has been the vigorous competition from other large players, such as Vodafone Group plc, the biggest wireless carrier in the world. In the Netherlands, where one in every 10 handsets sold is Internet-enabled, the Vodafone mobile Internet solution is selling four to five times better than iMode (as of December 2002). Vodafone Live is purported to be more interactive, and the MMS service on which it is based appeals to the youth segment, building on the early success of SMS services. The iMode product is claimed to be more complex for prospective consumers to comprehend, requiring the purchase of a handset, subscription, data bundle, and iMode content services (Telecom. paper, 2002).

NTT DoCoMo is also expected to launch iMode in the US market, purchasing a 16 per cent share of AT&T Wireless for 1 trillion yen in January 2001 in order to gain rapid market access (Shimbun, 2002). An iMode-type service, in partnership with AT&T Wireless, was originally planned for 2003. However, with the downturn in the telecommunications market and observed immaturity of the US market for wireless Internet services this has been postponed to December 2004. The services will be launched in San Francisco, Seattle, Dallas, and San Diego (Reuters, 2002a).

Developments in technology, service, and strategic partnerships

In terms of technological developments, DoCoMo continues to innovate in providing its iMode service. Although often referred to as the first, Japan was actually the second country to begin offering consumers access to 3G wireless networks in October 2001, second only to South Korea, where operators launched networks based on the CDMA2000 1x standard as early as October 2000 (SK Telecom launched in 2000, followed by KTF and LG Telecom in May 2001). DoCoMo's 3G network is based on the Wideband-Code Division Multiple Access standard. The service offered on the W-CDMA platform is called FOMA (Freedom of Mobile multimedia Access). These 3G networks will eventually allow transmission speeds of up to 2 Mbits/s – although typically only 384 Kbit/s – opening the way to advanced audio-visual capabilities.

Alongside network infrastructure, NTT DoCoMo is also working to further advance the iMode service platform. In January 2001, DoCoMo introduced 'iAppli' (short for 'information applications') for premium customers on its FOMA service. The new service, based on the Java programming language, provides a much higher level of sophistication in applications. These can be downloaded and stored, eliminating the need to continually connect to a Web site. Further, constantly changing information can be automatically updated at set times. This opens the door to such things as elaborate games and automatically updated stock quotes and weather reports.

Unfortunately for DoCoMo, the subscriptions to the FOMA service have been less than predicted. The full-year sales target from April 2002 to March 2003 was originally set at 1.38 million subscribers. However, in spite of features like face-to-face videophone calling and fast downloading from the Internet, frequent battery recharges, a limited coverage area, and high handset prices have put off users. Thus, with just less than 140,000 subscribers at the end of September 2002, the target has been cut considerably to 320,000 subscribers (Kawasaki, 2002). In December 2002, DoCoMo introduced 3G mobile phones with a much longer battery life – 170–180 hours in standby mode compared with 125 hours in its last model – and promised more improvements for 2003. Further, DoCoMo's 3G network now covers over 80 per cent of Japan's population and is set to reach 90 per cent by March 2003 (Reuters, 2002a).

During 2002–3, the FOMA service has come under pressure from other operators launching 3G offerings. KDDI Corp., the second largest operator in Japan, launched 3G services based on CDMA2000 in April 2002. Recent KDDI 3G phones typically last up to 200 hours or more between recharges and coverage exceeds that of DoCoMo. By December 2002, the operator had a subscription base of 4 million customers for 3G services (Reuters, 2002b). The third largest operator, J-Phone, owned largely by Vodafone Group, launched its 3G service on 20 December 2002, and aims to win one million users by the end of December 2003 (Global Wireless, 2002). J-Phone's offering, called Vodafone Global Standard, is based on the use of a dual terminal that can be used for both GSM and W-CDMA services, and a major feature of its service is the availability of global roaming on GSM networks (Kawasaki and Takebe, 2002).

Alongside technological developments, DoCoMo has sought to develop strategic alliances in a number of complementary areas, expanding the possibilities for platforms, devices, and services. Such alliances and projects include those with:

Ito Yokado Co. to provide a mobile cash card: the Mobile Cash Card service, planned for mid-2003, will enable customers to use an IYBank ATM with an iMode phone (currently the 504i series) by loading cash card data directly onto the phone (NE Asia, 2003). Customers type their password after transmitting the cash card data to an ATM through the infrared communication function. Then, they can settle their account in the same way as a traditional cash card. It is expected that a single phone can accommodate the cash cards of several different banks.

Sony Computer Entertainment for the Playstation game console: this project was aimed at creating network connectivity for the Playstation console by linking it to an iMode phone. This also allows iMode content to be accessed via the TV through the game console, and produces expanded functionality such as new data for old games (e.g. version patches or levels) and user uploads (e.g. competitive saved games).

Lawson Convenience for online retail: Lawson are one of the largest chains of convenience stores in Japan, with over 7500 shops nationwide. Lawson already have a sophisticated inventory system and 'Loppi' multimedia kiosks that allow shoppers to select products, print out a ticket with a barcode, and pay at the checkout counter to receive their purchase. The iLawson service allows users to make purchases via iMode handsets from a wide selection of books, CD, cosmetics, and other products. A purchase number is provided which is input into the Loppi kiosk to receive a ticket which can then be exchanged at the counter for the actual product. Customers select their preferred location for pickup and this information is used to build customer profiles. The more expansive version of the service, iConvenience, allows other companies to sell their products through Lawson stores (MacDonald, 2003).

Coca-Cola for new multimedia vending machines: as a visitor to Japan, it is easy to be surprised at the number of vending machines and range of products provided through them. In partnership with Coca-Cola, which has one million vending machines in Japan, a new service called Cmode was tested in August 2001, and from April 2002 been rolled out to many parts of Japan. Like most Japanese vending machines, Cmode machines accept coins and 1000-yen banknotes, and dispense a variety of differently flavoured soft drinks. In addition, Cmode machines also have a colour LCD, a camera with a barcode recognition system, a data connection to the iMode system, a numeric keypad, a Cmode button to activate certain functions, and a printer (Eurotechnology, 2002). Cmode has the capability of becoming an electronic wallet for consumers; after registering, users can deposit up to 5000 yen in their virtual account. By scanning a barcode on the iMode phone, users can buy drinks and other small products. The device can be also be used to sample and select digital content, such as ringing tones or graphics, which, once paid for by the digital wallet, can be downloaded onto a phone. Ticketing for amusement facilities, coupons, and maps are also a possibility using the built-in printer (MacDonald, 2003).

America Online and fixed-mobile convergence: Traditionally, as mentioned above, Japan has had a low penetration of wireline Internet connectivity. However, recognizing the limitations of mobile devices, the growth of wireline connections and the complementary nature of certain PC-based services, iMode has sought to bring convergence in service offerings. By investing in AOL Japan in 2000, DoCoMo created a new company, DoCoMo AOL, as a platform for launching new complementary services. Typical areas of complementary services include e-mail, instant messaging, and customer relationship management. In other areas, such as shopping, the PC provides a better platform for offerings.

Alongside, DoCoMo continues to build alliances with other partners, including those with banks to provide an online payment platform and car navigation systems vendors to create intelligent transport systems.

Conclusions

Based on the discussion of technology acceptance theory, it is perhaps unlikely that iMode's success will be emulated to the same extent or as easily in other markets. The conditions that have combined to create the fertile environment for the growth of iMode in Japan do not necessarily exist elsewhere. For example, the level of PC-based Internet access is already very high in the United States and Europe. Moreover, competition is more intense and technology fragmentation has been much higher and vertical integration lower. The NTT DoCoMo brand is less well known elsewhere, and cultural norms regarding conformity are also different. Nevertheless, given the high penetration of mobile phone use and the slow adoption of WAP, iMode could provide a compelling alternative for consumers in countries other than Japan. Key lessons to be learned from the success of iMode in Japan include the importance of a trusted, branded, useful, easy-to-use, holistic package of services, and the value of investment and leveraging of technological infrastructure such networks and handsets.

Acknowledgements

An earlier and much shorter version of this chapter first appeared as: Barnes, S. and Huff, S. (2003). Rising sun: iMode and the wireless Internet. *Communications of the ACM*, in press.

References

AU System (1999). WAP white paper. http://www.ausystem.com/, accessed 20 November 2000.

Ajzen, I. (1991). The theory of planned behavior. *Organizational Behavior and Human Decision Processes*, **50**, 179–211.

Asai, T. (2001). Transparent language. http://www.transparent.com/newsletter/japanese/2000/feb_00.htm, accessed 28 April 2001.

Business Week (2001). America next on DoCoMo's calling card. http://www.anywhereyougo.com/ayg/ayg/imode/Article.po?id=36786, accessed 31 January 2001.

Eurotechnology (2002). *The Unofficial Independent iMode FAQ*. Tokyo: Eurotechnology-Japan Corp.

Funk, J. (2000). *The Internet Market: Lessons from Japan's I-Mode System*. Kobe University, Japan: Unpublished White Paper.

Global Wireless (2002). J-Phone targets 1 million users by end-2003. http://www.globalwirelessnews.com/cgi-bin/news.pl? newsId=3758, accessed 5 December 2002.

Karahanna, E., Straub, D., and Chervany, N. (1999). Information technology adoption across time: a cross-sectional comparison of pre-adoption and post-adoption beliefs. *MIS Quarterly*, **23**, 183–207.

Kawasaki, S. (2002). High-flying NTT DoCoMo cuts sales targets for struggling FOMA service. http://neasia.nikkeibp.com/wcs/leaf? CID=onair/asabt/news/216370, accessed 12 November 2002.

Kawasaki, S. and Takebe, K. (2002). Int'l roaming is major feature of J-Phone's new 3G service; 'sha-mail' won't come with it. http://neasia.nikkeibp.com/wcs/leaf?CID=onair/asabt/news/220798, accessed 6 December 2002.

MacDonald, D.J. (2003). NTT DoCoMo's I-mode: developing win–win relationships for mobile commerce. In B.E. Mennecke and T. Strader, eds, *Mobile Commerce: Technology, Theory and Applications*. Idea Group Publishing, Hershey: pp. 1–25.

Matsunaga (2001). *The Birth of iMode*. Singapore: Chuang Yi Publishing.

Mobile Media Japan (2002). Japanese mobile Internet users. http://www.mobilemediajapan.com/, accessed 17 June 2002.

Nakada, G. (2001). I-Mode romps. http://www2.marketwatch.com/news/, accessed 5 March 2001.

NE Asia (2003). Cellular phones may replace wallets. http://neasia.nikkeibp.com/wcs/leaf?CID=onair/asabt/news/208033, accessed 24 September 2002.

Negishi, M. (2002). Mobile access to libraries: librarians and users experience for 'i-mode' applications in libraries. In *Proceedings of the 68th IFLA Council and General Conference*, Glasgow, August 18–24.

Nua (2002). Communications usage trend survey. http://www.nua.ie/, accessed 30 April 2002.

PMN (2002). Telefonica planning April I-Mode launch. http://www.pmn.co.uk/20021126telefonica.shtml, accessed 26 November 2002.

Rauhala, P. (2002). Comparision between Finnish and Japanese mobile markets. In *Proceedings of the International Workshop on Wireless Technology and Strategy in the Enterprise*, Berkeley, California, October 15–16.

Reuters (2002a). DoCoMo, AT&T to start 3G Service by Dec 2004. http://www.reuters.com/newsArticle.jhtml?type=topNews&storyID=1962131, accessed 30 December 2002.

Reuters (2002b). KDDI 3G subscribers top four million. http://www.reuters.com/newsArticle.jhtml?type=topNews&storyID=1872223, accessed 30 December 2002.

Rogers, E. (1995). *Diffusion of Innovations*. New York: Free Press.

Shimbun, N.K. (2002). DoCoMo stumbles at home, abroad. http://neasia.nikkeibp.com/wcs/leaf?CID=onair/asabt/news/210825, accessed 15 October 2002.

Sugiyama, Y. (2002). DoCoMo President Tachikawa anticipates 'taking new developments abroad in 2003'. http://neasia. nikkeibp.com/wcs/leaf?CID=onair/asabt/news/222595, accessed 17 December 2002.

TCAJ (Telecommunications Carrier Association of Japan) (2002). *Market Report*. Tokyo: TCAJ.

Tech Web (2001). I-Mode coming soon to phones worldwide? http://www.anywhereyougo.com/ayg/ayg/imode/Article.po?id=8648, accessed 12 January 2001.

Telecom.paper (2002). Vodafone live selling four times better than i-mode. http://www.telecom.paper.nl/site/news_ta.asp?type=abstract&id=23649&NR=150, accessed 23 December 2002.

Unstrung (2002). I-Mode's arrested development. http://www.unstrung.com/document.asp?doc_id=24640, accessed 19 November 2002.

van Miegham (2002). Bouygues Telecom to launch i-mode earlier than planned. http://www.europemedia.net/shownews.asp?ArticleID=12565, accessed 30 September 2002.

Venkatesh, V. and Davis, F. (2000). A theoretical extension of the technology acceptance model: four longitudinal field studies. *Management Science*, **46**, 186–204.

Wurster, T. and Evans, P. (2000). *Blown to Bits*. Boston: Harvard University Press.

Chapter 4

The Wireless Application Protocol as a platform for mobile services

Introduction

Until very recently the Internet and the mobile phone have appeared largely separate. However, since the mid-1990s, mobile technology providers have been working to bring convergence, enabling the wireless Internet. In 1995, Ericsson initiated a project to develop the Intelligent Terminal Transfer Protocol (ITTP) to provide a standard for value-added services in mobile networks. Similarly, in 1996, Unwired Planet launched the HDML and the HDTP, which, respectively, describe content/user interface and transaction protocols for wireless devices. Later, in 1997, Nokia introduced SMS and a language called Tagged Text Markup Language (TTML).

With a multitude of concepts there was substantial risk that the market could become fragmented. Therefore, all the major players agreed upon bringing forth a joint solution. The outcome was the Wireless Application Protocol (WAP) and the industry group involved is called the WAP Forum (www.wapforum.org) – a group with over 200 members dedicated to enabling sophisticated telephony and information services on handheld wireless devices (Logica, 2000). Indeed, WAP is now the most widely adopted wireless data protocol in the world among carriers and handset manufacturers (Saha *et al.*, 2001).

The implementation of WAP has provided both problems and opportunities. Nevertheless, it indicates an important starting point for the growth of the wireless Internet. Whether based on WAP or its successors, the wireless Internet promises to bring significant changes in the ways we live, work, and learn. During 2002, data were predicted to account for 20–30 per cent of all wireless network traffic, and by 2005, there could be more mobile phones connected to the Internet than PCs, accounting for around 30 per cent of all Internet traffic (Logica, 2000).

This chapter explores the strategic implications of WAP services. It provides a detailed analysis of the WAP service industry, including the role of customers, suppliers, rivalry, new entrants, and substitutes. The main focus for this chapter is B2C m-commerce – currently the fastest-growing sector (Bughin *et al.*, 2001). The chapter synthesizes and analyses some of the key strategic issues, before rounding off with some conclusions and recommendations.

The Wireless Application Protocol

WAP is a universal standard for bringing Internet-based content and advanced value-added services to wireless devices such as phones and PDAs. In order to integrate as seamlessly as possible with the Web, WAP sites are hosted on Web servers and use the same transmission protocol as Web sites, that is, Hypertext Transport Protocol (HTTP) (3G Lab, 2000). The most important difference between Web and WAP sites is the application environment. Whereas a Web site is coded mainly using HTML, WAP sites use WML, based on XML. WAP data flow between the Web server and a wireless device in both directions via a gateway that sits between the Internet and mobile networks. A wireless device will send a request for information to a server, and the server will respond by sending packets of data, which are formatted for display on a small screen by a piece of software in the wireless device called a micro-browser (Durlacher, 1999).

This section examines this architecture in more detail, touching on the role of the WAP gateway, the application protocol, and the programming environment. This sets the scene for later sections that assess the strategic implications of WAP.

The WAP gateway

An important part of the wireless communication infrastructure for WAP is the gateway – as shown in Figure 4.1. The WAP gateway provides the link between the Internet and mobile phone network. When a Web server receives a request from a wireless device for a WML file, it finds the file and then sends it out using HTTP – just like a Web page. The key difference between the WAP and Web model is that rather than sending the file intact straight through the Internet to the client device, the server sends it to a WAP gateway – typically a special piece of software. From the server's point of view, the WAP gateway is just like any other Internet Protocol (IP) address (3G Lab, 2000). Such gateways convert data received from a Web server via HTTP into a more compact and versatile format that is suited to the low bandwidths, high latencies, and intermittent connection of wireless telephony systems. Data are then sent on through the wireless network to the client mobile device using transmission procedures known as the WAP stack – similar to the procedures used in HTTP but optimized for communication across the airwaves rather than via wires. Finally, the client device's micro-browser receives the signals from the wireless network, processes them using the WAP stack, and updates its screen display (Logica, 2000).

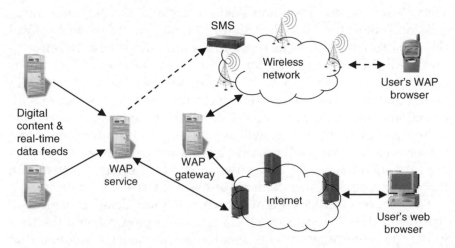

Figure 4.1   The role of gateways in the WAP architecture

Application protocol

As shown in Figure 4.2, the WAP protocol stack contains a number of important layers – the designs of which are broadly comparable to those of the wired Internet (3G Lab, 2000). In this section, we examine these in more detail, aiming to provide a deeper understanding of the nature of WAP.

Bearer service

At the physical level, WAP data are sent as streams of bits converted into radio signals. The physical system by which this transmission takes place is the domain of the network providers. Bearer services include:

- *Global System for Mobile (GSM)*: the prevailing mobile standard for around half of the world's mobile phone users, with speeds of up to 14.4 Kbits/s.
- *Short Message Service (SMS)*: a text messaging service based on GSM.
- *Unstructured Supplementary Services Data (USSD)* – an alternative method of messaging, based on GSM, offering real-time connection.
- *General Packet Radio Service (GPRS)*: a packet switched protocol based on GSM offering speeds of up to 115 Kbits/s.
- *Enhanced Data Rates for Global Evolution (EDGE)*: a higher bandwidth version of GPRS, allowing speeds of up to 385 Kbits/s.

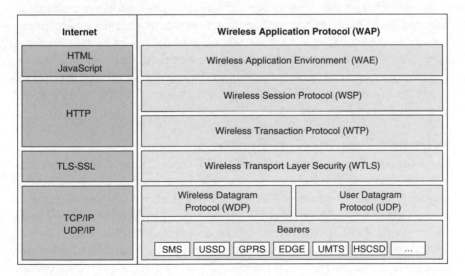

Figure 4.2 The architectures of WAP and the Internet compared

- High-Speed Circuit-Switched Data (HSCSD) – a circuit-switched protocol, based on GSM, that uses four radio channels simultaneously to achieve speeds of up to 57.6 Kbits/s.
- Universal Mobile Telephone System (UMTS) – the European 3G standard offering speeds of up to 2 Mbits/s.

Wireless Datagram Protocol

The Wireless Datagram Protocol (WDP) is the bottom layer of the WAP protocol stack. If WAP is used over a bearer supporting the User Datagram Protocol (UDP), the WDP layer is not needed (3G Lab, 2000). Essentially, as the most basic layers, WDP/UDP translate activity in the bearer service into a consistent data format for the higher layers of the WAP stack. In this way, WAP services can be implemented without consideration of the differences between individual bearer services. The WDP/UDP layer is broadly equivalent to the IP layer in the Internet model (see Figure 4.2).

Wireless Transport Layer Security

WTLS is an optional component of the WAP stack, derived from the Web's proven TLS and SSL standards. It provides the following key features (AU System, 1999):

- Data integrity: Ensuring that data exchanged between a mobile device and the server remain unchanged and uncorrupted.
- Privacy: Encryption technology scrambles data during transmission between the gateway and mobile device, protecting against interception.
- Authentication: checks are performed on the identity of both server and mobile device. If authenticity criteria are not met, access is withdrawn.
- Denial-of-service: to prevent services being disrupted by attempted attacks on provision, WTLS detects and rejects data that are replayed or not successfully verified.

Wireless Transaction Protocol

The Wireless Transaction Protocol (WTP) layer handles communication between the WAP client device, such as a mobile phone, and the gateway at the 'transaction' level: the level at which the client and server communicate with each other through individual requests and responses. WTP allows the client and server to check and, if necessary, recheck that wireless communication lines are strong before data transfer takes place. In essence, WTP is a streamlined protocol adapted to the technical limitations of wireless communications and small, handheld mobile terminals. It is broadly equivalent to the Transmission Control Protocol (TCP) in HTTP (Durlacher, 1999; 3G Lab, 2000).

Wireless Session Protocol

The Wireless Session Protocol (WSP) layer manages communication at the level of the client–server session. A session can comprise of one or more transactions. WSP is the interface between the Wireless Application Environment and the rest of the protocol stack. WSP is a binary version of HTTP 1.1 with a number of additions (AU System, 1999).

Developments to the WAP standard

Recent extensions of the first WAP standard have further developed capabilities in some key areas, including: image displays, public key infrastructure (PKI) and end-to-end security, messaging, and push technology. The newest version of WAP, version 2.0, supports many more advanced features for wireless information (see WAP Forum, 2002). For example, WAP 2.0 supports colour graphics, animation, file downloads, location-smart services, TCP/IP, and pop-up menus. In addition, some features of the Japanese iMode service have been integrated into WAP 2.0, particularly cHTML scripting via extensible HTML (xHTML), a language providing a framework for expandability and enhancement. The new xHTML protocol could bring thousands of existing Internet pages to mobiles, including iMode cHTML pages (Rauhala, 2002).

A strategic analysis of the provision of WAP services

Opportunities for IT-enabled competitive advantage vary widely from one company to another, just as the rules and intensity of competition vary widely from one industry to another. The complexity of the IT management challenge increases considerably when IT penetrates to the heart of a firm's (or industry's) strategy. Thus, in order to understand the impacts of the wireless Internet and WAP we need a comprehensive strategic framework. Porter (1980) provides such as framework, arguing that economic and competitive pressures in an industry segment – such as the WAP service industry – are the result

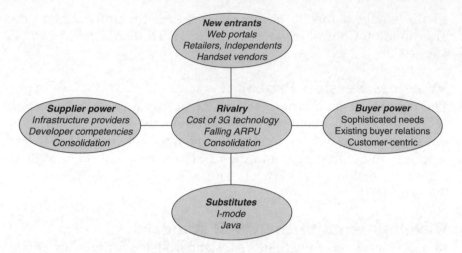

Figure 4.3   Strategic analysis of the WAP service industry

of five basic forces: (a) positioning of traditional intra-industry rivals; (b) threat of new entrants into the industry segment; (c) threat of substitute products or services; (d) bargaining power of buyers; and (e) bargaining power of suppliers.

This section aims to provide an analysis of the WAP service sector from a strategic viewpoint. The purpose of this analysis is to provide some understanding of the key forces impacting on the ability of WAP services to succeed in provision of mobile Internet services. The results of such an analysis have distinct implications for practitioners. Figure 4.3 shows Porter's framework, highlighting some of the key elements for each of the five forces. As we can see, there is strength in each of the forces, emphasizing a very strong degree of competition in the provision of WAP services. Let us examine the framework in more detail.

Rivalry

The network operators are powerful players in the industry that provides WAP services. Traditionally, there has been a reasonably high concentration of players in mobile telecommunications. Recently, rivalry has been exacerbated by developments in service pricing and future service provision. With the implementation of 3G transmission technologies on the horizon, network operators have

been clambering for licenses to provide services. The next generation of technologies promise transmission speeds of up to 2 Mbits/s,
opening the door to a raft of high bandwidth services and multimedia. However, the cost of access to such applications – in terms of the
frequency licensing arrangements – have not been cheap; in the
United Kingdom, the cost of 3G License B purchase soared to over
£20 billion, and similar figures were seen in other European countries (e.g. Germany). Such costs will inevitably need to be passed on
to the consumer.

On the other side of the coin, the mobile market has been
squeezed in terms of consumer pricing arrangements. As network
operators have sought to increase the volume of mobile telecommunications and to provide differentiated packages to the customer,
profit margins have fallen. Generally speaking, the Average
Revenue Per User (ARPU) has declined steadily over the last 10
years and is now an estimated 77 per cent lower than that of 1990
(Barnett *et al.*, 2000).

The industry response has been a global consolidation as operators try to deal with their high up-front investments for 3G and
decreasing ARPU. This consolidation trend, along with control of
access and direct ownership of the customer make the operators the
most powerful players in WAP services (WireFree-Solutions, 2000b).
However, legal issues reduce the power of operators to restrict
access to content, and a decrease in barriers to entry increases competition from mobile Internet content providers.

All of the other forces in Porter's framework – new entrants, substitutes, buyers and suppliers – further contribute directly to rivalry.
These are now explored in turn.

New entrants

The predicted revenues from wireless data services are enormous and
have provided an attractive impetus to the entry of new players into
the industry. However, incumbent operators, suffering competitive
pressures, have used their control of the network infrastructure to try
and lock-in potential value; by presetting subscribers' telephones to
make themselves the default Internet access provider and blocking
unauthorized services, operators have the opportunity both to charge
application providers for access to their subscriber base and to build
their own branded services (Barnett *et al.*, 2000). Nevertheless, where

an industry is driven by consumer choice and varied access to services, such a strategy may not prove to be effective in retaining customers into the longer term.

The key B2C market makers on the mobile Internet are mobile portals (or m-portals) – revenues of which are predicted to be $42 billion by 2005 (Ovum, 2000). In Europe alone, the market is expected to be worth $10 billion by 2005 (Bughin *et al.*, 2001). Literally, the word 'portal' means a doorway or gate; mobile portals are high-level information and service aggregators or intermediaries (Chircu and Kauffman, 2001) that provide a powerful role in access to the mobile Internet. Their main aim is the provision of a range of content and services tailored to the needs of the customer, including:

- communication, for example, e-mail, voice mail, and messaging;
- personalized content and alerts, for example, news, sports, weather, stock prices, and betting;
- personal information management (PIM), for example, 'filofax' functions; and
- location-specific information, for example, traffic reports, nearest bank or ATM, film listings, hotels and restaurant bookings.

As such, mobile portals are usually characterized by a much greater degree of customization and personalization than standard Web-based portals in order to suit the habits of the consumer (Durlacher, 2000). The mobile portal must be suitably tailored to the user's needs so as to present the right information at the right time on the small-screen WAP device.

The mobile portal market has undergone significant expansion in anticipation of market growth. More than 200 WAP portals have been launched in Europe alone since Autumn 1999 (Bughin *et al.*, 2001). Players have attempted to build on existing brands, competencies, and customer relationships to develop a subscriber base. Key players have been:

- Mobile operators. Portals include Genie (BT), Zed (Sonera), and MyDof (Telia).
- Technology vendors. For example, Nokia, Ericsson, Palm, and Motorola have all developed portal services.
- Traditional Web portals. Including offerings from Yahoo!, AOL, and Excite.
- Retail outlets. For example, the Mviva portal is owned 85 per cent by Carphone Warehouse.

- Random new entrants. Including portal services from banks, for example, Barclays, and mass-media companies, for example, the Vizzavi portal is half owned by Vivendi.
- New independents, including Iobox, Room33, and Quios.

Given time, one might expect the portal market to consolidate, although the potential role of niche players appears much greater than that of the traditional Web portal market.

Substitutes

WAP adoption by consumers is both patchy and limited. As of July 2001, the use of WAP phones has been disappointingly low; just 6 per cent of Finnish and US mobile phone users access the Internet using their phones, compared with only 10 per cent in the United Kingdom and 16 per cent in Germany (eMarketer, 2001). Predictions are much better for some parts of the Asia-Pacific (Dataquest, 2000). In Japan, the success of WAP services has been greatest, with 11 million subscribers to the EZWeb WAP service in September 2002 (Mobile Media Japan, 2002).

In most countries, the expectations of consumers have not been met and WAP has been considerably oversold. Part of the problem is the limitation of technology and the non-subtractive nature of services; WAP is not a replacement for the wired Internet and involves an important trade-off between richness and reach in providing data services (Wurster and Evans, 2000). Furthermore, whilst proponents argue that WAP is scalable and extensible enough to endure (Leavitt, 2000), many see WAP as a stopgap until 3G phones. In particular, critics point to the primitive nature of WAP, which is too closely aligned to the current generation of mobile phones, and the possible control of material by cellular operating companies, which will stifle creativity (Goodman, 2000). Key problems include security (Korpela, 1999), some of which have been addressed by recent versions of WAP, the high cost (until all networks become packet-switched and the pricing model changes), and limited infrastructure (from networks and devices) (Barnes *et al.*, 2001).

In addition to WAP, much attention is now being drawn to the HTML-based iMode standard in Japan. The growth and success of iMode provides considerable food for thought for WAP proponents. Launched in February 1999, iMode has a subscriber growth rate of nearly 1 million per month, standing at 35 million in September 2002 (Mobile Media Japan, 2002). This is more than three times more than the competing WAP service, EZWeb. NTT DoCoMo, the owners of the iMode brand and service, are now planning to 'export' this model to the United States and Europe. Through a strategy of partnering, NTT DoCoMo hopes to emulate its earlier success (Associated Press, 2001; Business Week, 2001).

Analysts put this success down to a number of reasons including technological investment, market dominance, vertical integration in technology development, and the low penetration of expensive wired Internet (Funk, 2000; Kramer and Simpson, 1999; WireFree-Solutions, 2000a).

Clearly, the development of iMode is very different to WAP. In some senses the Japanese iMode example is unique and perhaps unlikely to be emulated in the United States (Diercks and Skedd, 2000) and elsewhere. However, there appear to be some lessons that can be gleaned. iMode is very definitely a brand and stands for key concepts like simplicity, functionality, and meeting consumer needs (see chapter 3). In this respect, WAP has some way to go to catch up with iMode; WAP is a bundle of technologies and protocols, which on its own does not deliver value to the end-user.

Another possible alternative to WAP are standards based on Java – a 'write once, run anywhere' programming language – to provide a full application execution environment. These include Java 2 Micro Edition (J2ME) (Newsbytes, 2001) and the Mobile Station Application Execution Environment (MExE) (Durlacher, 1999). These standards are primarily aimed at the next generation of powerful smart-phones. MExE, for example, incorporates some advanced features to provide intelligent customer menus, voice recognition, and softkeys, as well as to facilitate intelligent network services.

Customers

Although the network operators still dominate the wireless market as a key intermediary, the signs are that this situation will change very quickly and may – to some extent – mirror the business model of ISPs in the traditional Internet market (Mobilocity, 2000). The key driver here is service provision; the operators, in order to increase their ARPU, have to provide services that increase the customer's willingness to pay. Whereas there are currently very few services linked to cellular telephone companies (cellcos), estimates suggest that by 2004, around 75 per cent of wireless revenues will be from the provision of services, including those based on WAP (KPMG, 2000).

The convergence of mobile telecommunications and the Internet leads to more personalization and customer empowerment (Arthur D. Little, 2000; Barnett *et al.*, 2000). Although segmentation is important, for example, focusing on specific market segments with designated products or specific high-quality products for various segments, there is a strong recognition that one size does not fit all on the wireless Internet (Arthur D. Little, 2000). After all, a customer who is not fully satisfied with a WAP service can move to another service provider immediately. Three key features enable the personalization of wireless Internet: the 'always at hand' nature of the mobile phone; the unique identifying nature of the phone; and the ability to detect a user's location (Barnett *et al.*, 2000). Using 'intelligent' personalization tools, such information can be used to enhance the richness of the user's service experience, anticipating customer needs.

Presently, the dominant business model for WAP service provision involves mobile operators aggregating content and services from third-party partners and providing these services directly to their subscribers. This situation is shown in Figure 4.4, where m-businesses and portals are obliged to reach customers through proprietary networks. However, as the diffusion of WAP accelerates and consumers begin demanding services independent of the wireless carrier, a model similar to that of ISPs is likely to emerge (as indicated in Figure 4.5); under this model, wireless service subscribers will have access to any mobile site, and the open-access system will spur companies' development of their m-commerce presence (Mobilocity, 2000). However, new entrants will be severely

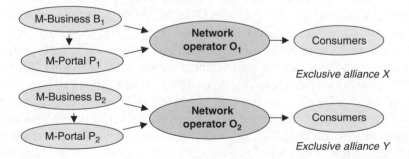

Figure 4.4 The closed-operator model (adapted from Mobilocity, 2000)

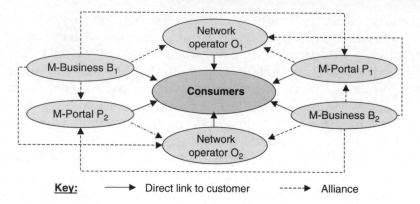

Figure 4.5   The customer-centric model (adapted from Mobilocity 2000)

challenged by incumbents who have built strong customer relation-ships. Unlike the Web – where customers traditionally maintain multiple accounts – companies in the WAP service industry will be highly dependent on network operators and first movers who con-trol billing relationships. In this sense, the customer-centric model will not prevent companies from creating partnership consortiums to provide the entire range of value-added services to consumers (Barnett *et al.*, 2000).

Suppliers

Suppliers of handheld mobile devices are one key group that have a strong grip on the WAP service industry. In the smartphone market, as in the PDA market, the brand and model are the most important part of the purchase decision; the service provider or network provider is less important (Peter D. Hart, 2000). A stable oligopoly of four smartphone suppliers set prices to sell what they can produce; there is no omnipotent force pressuring prices, and little evidence that low-cost strategies win market share (Kramer and Simpson, 1999). Players such as Nokia have proven that barriers to entry in the handset market are substantial, with the cost of branding, produc-tion capacity, and research and development deflecting considerable competition and making high margins sustainable. The next wave of consolidation in the wireless industry will most likely involve hand-set vendors strengthening their position prior to the new wave of sophisticated wireless services (Kramer and Simpson, 1999). Simple strategies of 'safety in numbers' will not address the deep impacts of deploying next-generation networks and services; access to leading-edge competencies in software and services is likely to be more important than being the largest supplier of a given element of the network.

Mobile network operators – such as KPN Mobile, Sonera, and Vodafone – are an important part of the transport process. Notwithstanding, these players are now leveraging their infrastructure advantages in transport to enable movement along the value chain towards mobile services, delivery support, and market making. Typically, these operators control the billing relationships and SIM cards on WAP phones and are ideally positioned to become mobile Internet service providers (MISPs) or portals, thereby establishing a transport pipeline for content services (Durlacher, 1999).

Apart from the infrastructure suppliers who are driving technological progress, a host of other suppliers are important – such as those who handle financial transactions, software application developers, content packagers, and content providers. The simple value chain that is mostly controlled by the network operator and heavily influenced by handset vendors is being transformed into a complex value network where alliances play a key role. In this new digital economy, consumer online services demand that diverse inputs must be combined to create and deliver value. No single industry alone has what it takes to establish the m-commerce economy (Barnes, 2002).

Discussion and analysis

The WAP service industry is in a state of flux. Driven by this new platform of value-added services, rivalry has begun to develop. Pressured by falling revenues, operators are seeking to build on important relationships with mobile customers – such as billing – to extend their portfolio of offerings. However, the demands for services from customers are unlikely to be met from a sole company; with the opening of WAP channels to the customer, partnership is a key trend in an area where pressure is mounting from players entering the market either directly or from adjacent competencies. In other parts of the value chain, the power of suppliers – such as handset vendors – is also distorting the market.

Figure 4.6 provides a simple framework for visualizing some of the key strategies in the WAP service industry. In particular, this shows how strategies are likely to change over time, driven by the increasing trend towards an open, customer-centric industry model and full service provision. The matrix has two axes: market focus and channel access. In the framework, market focus can either be broad, as the WAP service provider aims to be a portal, or niche, as the WAP service provider aims to target a specific segment. In terms of channel access, this can either be closed, where control falls to the network operator as an intermediary, or open, where access to the customer is direct and partnerships are likely to play a role in the provision of a range of services.

In the early days of WAP, services were content-focused, as indicated by the leftmost cells on the grid; the provision of WAP services is based more on a supplier 'push' than a customer 'pull'. Typically, the network operator has played the role of content enabler, providing a range of selected content services to its subscribers. It is the controlling faction in exclusive alliances. Such players include KDDI/au, Vodafone, and Sonera. Other companies – content providers – supply focused, niche-oriented digital content to the network operator portal. Examples include Kizoom, the travel information provider, and BBC News Online, via its WAP news site.

As the WAP service market becomes more open, the user is likely to become the key focus; in this new era, WAP services become demand-led by the ever-sophisticated needs of the consumer. As the

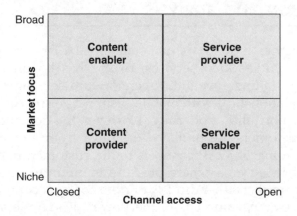

Figure 4.6   Strategic framework for WAP service provision

channels to the consumer become more accessible, other players will enter the increasingly lucrative portal market, attempting to gain a share of increasing service revenues. In a market driven by personalization and consumer choice, alliances provide an important way to give the full range of consumer-demanded services. Participants in such alliances – service enablers – are an integral part of service offerings. Those who provide the 'front-end' of these offerings – service providers – are far less dominant than in the closed-channel era. Nonetheless, network operators are still likely to be central players in the early stages of open-channel access due their knowledge of the customer and pre-existing relationships – particularly via billing.

Conclusions

In spite of some misgivings, the wireless Internet is now firmly on the map, aided by the platform provided by WAP. The WAP standard provides the first steps in a path towards mobile Internet, taking stock of the current limitations of wireless networks and mobile devices. As Porter's framework has demonstrated, WAP creates an interesting and powerful set of dynamics for the industry of mobile Internet service provision, with competition and collaboration coming from a variety of avenues. Mergers and acquisitions have been rife as players in wireless, IT, and media industries have attempted to position themselves, under the increasing threat of competition. As well as the transformation of incumbents, the lure of service revenues and the drive to serve the customer brings many new players from adjacent and even unrelated markets. In order to understand some of the strategic implications of WAP for such companies, this chapter has provided an original matrix framework to chart some of the key market strategies.

The above analysis provides some important implications for practitioners. The provision of WAP services, although currently dominated by key players such as operators and infrastructure providers, must become more open and inclusive in the future in order to succeed in core markets such as the United States and Europe. As early apathy to WAP services has shown, successful offerings must be demand-led rather than supply-driven; the penetration of WAP services has so far been limited, and initial high expectations of the wireless Internet have not converged with the reality. Successful WAP offerings are likely to be those combining content, infrastructure, and services in a seamless way, attempting to be relevant and personal to the mobile phone user. The core consumer market for mobile Internet services is likely to be users under age 35, whose trade-off between reach and richness has proved most favourable (Wurster and Evans, 2000). In Japan, a massive 72 per cent of mobile phone users access the Internet via their phones, largely on the iMode platform (eMarketer, 2001). Although market conditions are very different in the United States and Europe, iMode does provide some important lessons, including the importance of a trusted, branded, holistic package of services, and substantial investment in and leveraging of superior technological infrastructure, such as networks and handsets. Such ideas provide significant food for thought for industry players involved in WAP service provision.

Whether WAP continues to thrive into the medium-term is uncertain. The implementation of the next generation of transmission technologies will enable a new breed of high bandwidth mobile networking that will stretch the abilities of WAP. Alongside, mobile devices are becoming more powerful – combining the capabilities of a mobile phone and small computer into a PDA. Whether WAP is extensible enough to cope with the possibility of rich multimedia and 'always-on' connection remains to be seen. WAP will always exist as a technology alternative, but the strengths of other application protocols such as Java-based MExE and HTML-based iMode provide attractive replacements. Such replacements are built for a world where complex interactivity is paramount. Notwithstanding, in the absence of more advanced infrastructure WAP provides the de facto standard for the wireless Internet; WAP will most likely endure into the short term, spearheading initial attempts at wireless data services for B2C markets. During 2001–2, growth of WAP services and sales of WAP phones in some parts of Europe and Asia were surprisingly buoyant, although this was less so in North America.

Acknowledgements

This chapter first appeared as Barnes, S.J. (2002). Provision of services via the Wireless Application Protocol: a strategic perspective. *Electronic Markets*, **12** (1), 1–8.

References

Arthur D. Little (2000). Serving the mobile customer. http://www.arthurdlittle.com/ebusiness/ebusiness.html, accessed 21 November 2000.

Associated Press (2001). NTT DoCoMo to offer iMode in Europe later this year. http://www.anywhereyougo.com/ayg/ayg/wireless/Article.po?id=1034, accessed 18 January 2001.

AU System (1999). WAP white paper. http://www.ausystem.com/, accessed 15 February 2000.

Barnes, S.J. (2002). The mobile commerce value chain: analysis and future developments. *International Journal of Information Management*, **22**, 91–108.

Barnes, S., Liu, K., and Vidgen, R. (2001). Evaluating WAP news sites: the WebQual/m approach. In *Proceedings of the European Conference on Information Systems*, Bled, Slovenia, June.

Barnett, N., Hodges, S., and Wilshire, M. (2000). M-commerce: an operator's manual. *The McKinsey Quarterly*, No. 3, 163–173.

Bughin, J., Lind, F., Stenius, P., and Wilshire, M. (2001). Mobile portals: mobilize for scale. *The McKinsey Quarterly*, No. 2, 118–127.

Business Week (2001). America next on DoCoMo's calling card. http://www.anywhereyougo.com/ayg/ayg/imode/Article.po?id=36786, accessed 15 January 2001.

Chircu, A. and Kauffman, R. (2001). Digital intermediation in electronic commerce – the eBay model. In S. Barnes and B. Hunt, eds, *Electronic Commerce and Virtual Business*, Oxford: Butterworth-Heinemann, pp. 45–64.

Dataquest (2000). Asia-Pacific mobile Internet service dominated by WAP in first quarter of 2000. http://gartner6.gartnerweb.com/dq/static/about/press/pr-b09182000.html, accessed 18 September 2000.

Dierks, B. and Skedd, K. (2000). Global demand for wireless Internet on the upswing – carriers must structure services towards specific markets. http://www.instat.com/pr/2000/md2004md_pr.htm, accessed 18 October 2000.

Durlacher (1999). Mobile commerce report. http://www.durlacher.com/research/, accessed 15 January 2000.

Durlacher (2000). Internet portals. http://www.durlacher.com/research/, accessed 15 June 2000.

EMarketer (2001). Wireless Web growing around the world. http://www.nua.ie/surveys/index.cgi?f=VS&art_id=905357175 &rel=true, accessed 10 September 2001.

Funk, J. (2000). *The Internet Market: Lessons from Japan's I-mode System.* Unpublished White Paper, Kobe University, Japan.

3G Lab (2000). *Your Pocket Guide to the Mobile Internet.* Cambridge: 3G Lab Limited.

Goodman, D. (2000). The wireless Internet: promises and challenges. *IEEE Computer*, **33**, 36–41.

Korpela, T. (1999). *White Paper of Sonera Security Foundation v. 1.0.* Helsinki: Sonera Solutions.

KPMG (2000). *Wireless Report.* London: KPMG.

Kramer, R. and Simpson, B. (1999). *Wireless Wave II: the Data Wave Unplugged.* London: Goldman Sachs.

Leavitt, N. (2000). Will WAP deliver the wireless Internet? *IEEE Computer*, **33**, 16–20.

Logica (2000). *The Mobile Internet Challenge.* London: Logica Telecoms.

Mobile Media Japan (2002). Japanese mobile Net users. http:// www.mobilemediajapan.com/, accessed 23 December 2002.

Mobilocity (2000). Seizing the m-commerce opportunity: strategies for success on the mobile Internet. http://www.mobilocity.net/, accessed 28 May 2000.

Newsbytes (2001). RIM founder lauds J2ME as common platform for wireless. http://www.ayg.com/j2me/Article.po?id=1575716, accessed 12 July 2001.

Ovum (2000). Wireless portal revenues to top USD42 Bn by 2005. http://www.nua.ie/surveys/index.cgi?f=VS&art_id=905355888 &rel=true, accessed 5 July 2000.

Peter D. Hart (2000). *The Wireless Marketplace in 2000.* Washington DC: Peter D. Hart Research Associates.

Porter, M. (1980). *Competitive Strategy.* New York: Free Press.

Rauhala, P. (2002). Comparison between Finnish and Japanese mobile markets. Presentation to the *International Technology and Strategy Forum*, University of California at Berkeley, October.

Saha, S., Jamtgaard, M., and Villasenor, J. (2001). Bringing the wireless Internet to mobile devices. *IEEE Computer*, **33**, 54–58.

WAP Forum (2002). Wireless Application Protocol 2.0 – technical white paper. http://www.wapforum.org/, accessed 23 December 2002.

WireFree-Solutions (2000a). WAP vs. I-Mode – let battle commence. http://www.wirefree-solutions.com/, accessed 30 June 2000.

WireFree-Solutions (2000b). An overview of the wireless Internet with WAP. http://www.wirefree-solutions.com/, accessed 20 June 2000.

Wurster, T. and Evans, P. (2000). *Blown to Bits*. Boston: Harvard University Press.

Chapter 5

Under the skin: short-range embedded wireless technology

Introduction

Aside from products that are specifically wireless, such as PDAs and smartphones, there are many more that are wireless-enabled via the inclusion of a wireless communication unit. These units are known as 'embedded' wireless devices, since they are included 'under the skin' of the product. Typically, they allow short-range data communications, and could perform functions of cable replacement in an office environment. Industry sources forecast that by 2003, the number of wireless computing devices will exceed the population of our planet – estimated at 6 billion. This figure is predicted to consist of a mixture of: around 300 million PDAs; approximately 2 billion consumer electronic devices such as wireless phones, pagers, and set top boxes; and 5 billion additional everyday devices like vending machines, refrigerators, and washing machines embedded with chips connected the Internet (IBM Pervasive Computing, 2000).

This chapter reflects on this important issue. As a foundation, it critically examines the key enabling technologies for commercial applications of short-range embedded wireless devices. Such advances are many and varied and have tended to be shrouded in technical complexity. This chapter consolidates these into an accessible form. It explores the potential impact of embedded wireless applications in a number of key areas: home, work, travel-related, and those that are publicly accessible. It draws on some state-of-the-art case examples and the current opportunities and problems in developing such applications. Finally, the chapter concludes with reflections on the future of embedded wireless technologies.

Embedded wireless technologies for short-range communication

Computers and related electronic devices have become an increasingly important part of modern life – in the home, at school, and at work. Such technologies are ubiquitous in developed economies and are rapidly penetrating many developing economies. An estimated 1.39 billion people have PCs, including 55 per cent of those in Asia, 51 per cent in North America, 39 per cent in Western Europe, and 27 per cent in Latin America (Roper Starch, 2000). In mature markets such as North America and Europe, usage is so significant that saturation is beginning to occur (IDC Research, 2000).

As the number of computing and telecommunications devices used at home and in business has increased, so has the complexity and difficulty of physical wiring. The usual solution is to connect the devices with a cable to make file transfer and synchronization possible. File transfer is needed, for example, to move documents between devices, such as PCs and PDAs. Synchronization is also important to achieve commonality of information, such as calendars and event-based applications. The cable solution is often complicated, since it may require a cable specific to the devices being connected as well as configuration software. Another solution, infrared, eliminates cable but requires line-of-sight.

To solve some of these problems, various other technologies have been created to provide the means for short-range radio connection via embedded wireless devices (see Table 5.1). Perhaps the best known of these is Bluetooth – the cooperative effort of a number of companies working for a cheap, simple, and low power-consuming solution with broad market support. The next section summarizes the key technologies available, examining their specifications and characteristics.

Table 5.1 Competing technologies for short-range wireless communication

Technology	Payload[a]	Range[b]
Bluetooth	723 Kbits/s in 2.4 GHz band	10–100 m[c]
IrDA (Infrared Data Association)	4 Mbits/s for IrDA data 75 Kbits/s for IrDA control	2 m for IrDA data 5 m for IrDA control
IEEE 802.11 of (Institute Electrical and Electronics Engineers)	11 Mbits/s for 2.4 GHz band (802.11b) 54 Mbits/s for 5 GHz band (802.11a)	25–600 m[c]
DECT (Digital Enhanced Cordless Telecommunications)	736 Kbits/s in 1.88 GHz band	300 m to 25 km
HomeRF (Home Radio Frequency)	1 Mbits/s in 2.4 GHz band	Approx. 40 m
UWB (Ultra-Wideband Radio)	1.25 Mbits/s	Approx. 70 m

Notes: [a]Measured in Kilobits/second (Kbits/s) and Megabits/second (Mbits/s). Frequency given in Gigahertz (GHz)
[b]Measured in metres (m) and kilometres (km)
[c]Implies a graceful degradation of data rates

Bluetooth

Bluetooth is a data communication specification designed to enable wireless communication between small, mobile devices. The name Bluetooth is derived from a Danish Viking King, Harald Blåtand (Bluetooth), who lived in the late tenth century and united and controlled Denmark and Norway. The standard was developed by the Bluetooth Special Interest Group (SIG); an industry group consisting of leaders in the telecommunications and computing industries that are driving development of the technology and bringing it to market. The group consists of around 2000 members and includes 3Com, Ericsson, IBM, Intel, Lucent Technologies, Microsoft, Motorola, Nokia, and Toshiba (AnyWhereYouGo, 2000). The key benefits of Bluetooth are the minimal hardware dimensions, low price of components, and low power consumption (AU System, 2000). Moreover, the diversity of product offerings from companies in the SIG that support Bluetooth (e.g. mobile phones, PDAs, computers, computer hardware, and software) creates a unique market position. Bluetooth uses a spread-spectrum, frequency-hopping technique to protect against interference (e.g. from a microwave oven), allowing 1600 hops per second. In addition, each network may control up to eight devices (Intercai Mondiale, 2000). The cost of Bluetooth chips is currently approximately $9, but it is expected to fall significantly to around $4 by 2005, as the chips become more frequently embedded in all manner of devices (Shim, 2002). Bluetooth is expected to become the preferred standard for embedded wireless devices, particularly in the consumer market.

Infrared Data Association

The most established player in the cable replacement market is IrDA (Infrared Data Association); an infrared standard providing wireless solutions for serial data connection between, for example, notebook PCs and controlling devices. The technique is well known in the market and around 100 million ports have been installed (Lambert, 2000). IrDA is relatively simple to configure and use, given suitable software support, although it has had some problems through IrDA manufacturers that have created incompatible devices (AU System, 2000). IrDA for data transfer has the advantage of a high throughput (payload) compared to other standards such as Bluetooth. However, IrDA is limited to two parties, point-to-point connection and needs line-of-sight for its infrared beam.

Institute of Electrical and Electronics Engineers 802.11 standard

The primary competitor for Bluetooth in the market for wireless enterprise data is the Institute of Electrical and Electronics Engineers (IEEE) 802.11 group of standards. The 802.11b standard is typically referred to as WiFi (wireless fidelity) or wireless LAN (WLAN). Like Bluetooth, some implementations of this standard use frequency-hopping technology, although not to the same extent as Bluetooth. In addition, IEEE 802.11 enjoys a number of benefits, not least its very high transmission capacity and large number of possible simultaneous users. At the end of 2002, this standard was fast becoming the preferred method of data connectivity for enterprise wireless applications based on the use of connectivity 'hot-spots'. The cost of WiFi chips is typically in the order of $15–20 at the end of 2002 (Merritt, 2002).

Digital Enhanced Cordless Telecommunications

The Digital Enhanced Cordless Telecommunications (DECT) standard is one of the oldest, developed more than 10 years ago by the European Telecommunications Standards Institute (ETSI). It is promoted by the DECT Forum and used in over 100 countries for data and voice services (Lambert, 2000). More recently, it has been adopted as a technology for 3G wireless transmission standards, based on IMT-2000 (International Mobile Telecommunications 2000 standard). The current DECT derivatives operating in the unlicensed ISM (Industrial, Scientific and Medical) 2.4 GHz band – the frequencies of Bluetooth and IEEE 802.11 – are proprietary with limited functionality. In particular, they tend to focus on advanced telephony systems for the home or small office/home office (SOHO) market, with typical products available from Siemens and Ericsson. Nevertheless, the throughput is comparable to Bluetooth and the possible range is the most extensive of the technologies listed in Table 5.1.

Home Radio Frequency

Home Radio Frequency (HomeRF) is a PC-centric wireless solution aimed particularly at home networking (Lambert, 2000). A consortium of, among others, Microsoft, Intel, Motorola, and Compaq, developed SWAP (Shared Wireless Access Protocol) for data and voice wireless communication in HomeRF. HomeRF is based on the DECT concept, with influences from IEEE 802.11, and operates in the same frequency band. There are many similarities with Bluetooth in areas such as price per unit, range, and transmission power (HomeRF, 2000); the major differences are that HomeRF can handle more devices (up to 127 per net) and uses just 50 frequency hops per second. During 2001, there is evidence that interest in HomeRF is beginning to diminish in favour of other solutions.

Ultra-Wideband Radio

One very recent technological development is Ultra-Wideband Radio (UWB) – a concept similar to that of radar. In UWB, short pulses are transmitted in a broad frequency range, and the information is modulated by the pulses' time and frequency (AU System, 2000). The technique is not yet fully developed, but it could pose a threat to standards such as Bluetooth due to its higher transmission capacity and very low power consumption (just 0.5 mW).

Commercial applications of short-range embedded wireless

Clearly, the kinds of technologies outlined above have tremendous commercial potential. This section explores some of the main applications of short-range embedded wireless technologies that are currently under development. As a starting point, these are examined using a simple categorization, as presented in Figure 5.1.

Applications of Bluetooth and other related technologies are typically aimed at domestic or home use, use in the office or other workplace, applications for public consumption, or travel purposes. Typically, however, many applications fall into more than one category; for example, the smart car with a wireless network can become an office in-transit. Therefore, there is a degree of mobile convergence, with the possibility of some advanced devices that

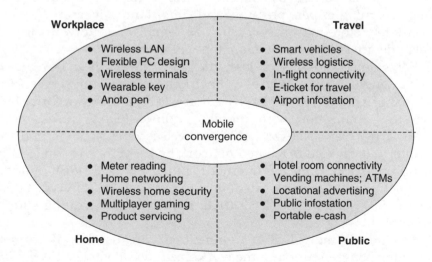

Figure 5.1 Applications of short-range embedded wireless – examples

could be developed to fulfil a large number of activities in various categories shown in Figure 5.1. Each of the key applications in Figure 5.1 is now explored in more detail.

Applications in the workplace

Within the office or other workplace, technologies such as Bluetooth create a new era of product design. Specifically, it becomes possible to remove all cabling except the power supply to electronic devices; for example, the telephone, keyboard, loudspeakers, PC screen, and the PC itself can all be connected using short-range embedded wireless technology (AU System, 2000). The removal of signalling cables can lead to new ways of furnishing an office, as the central processing unit (CPU) no longer needs to be next to the monitor and keyboard.

In addition, laptops will be lighter, since they do not require as many connecting ports. Toshiba and IBM are both beginning to embed Bluetooth and WiFi antennae around the display screen of notebook PCs. Taken to the extreme, advanced control devices such as the Anoto pen could allow incredible flexibility (Anoto, 2000). This pen – appearing like an ordinary ballpoint – is embedded with a digital camera, advanced image processing, and Bluetooth connectivity. It works by being passed over a piece of paper printed with a special pattern. By collecting information on x–y coordinates, angle, turning of pen, pressure, and time-stamps, it can be used to accurately store information (such as e-mails) and forward it to a relay device (such as a mobile phone, PDA, or PC).

Some devices for the wireless office have already been developed. For example, i-data produces a Bluetooth-enabled print server – the PlusCom Xpress PRO 10/100 BT – that has been designed for instant wireless connection with mobile phones, PDAs, and laptops (Newsbytes News Network, 2000). As well as wireless connection, the printer is connected to a wired network by Fast Ethernet connection, including support for the Internet Printing Protocol (IPP). In addition, RTX Telecom produces a dongle to plug into PC Universal Serial Bus (USB) ports – the RTX 9201 Bluetooth USB – creating simple wireless PC connectivity.

Beyond the individual PC, embedded wireless also allows networking between people in the workplace – via a wireless LAN. This 'workplace' could be any location with workers having SRW-enabled devices. For example, ad-hoc groups could form in a coffee shop or at a conference to work collaboratively on a document (AU System, 2000). Such connectivity has been provided in the USS McFaul – a state-of-the-art American Navy destroyer (Seybold, 2000). On McFaul, PDAs – specifically Palm Vs – are the basis of a wireless data network that also includes an infrared LAN and Aether Systems server software. The Aether software monitors the network and knows where and when the Palm Vs are being used, and it bridges the ship's Microsoft Exchange-based intranet to the hand-held devices by converting the data into a Palm-readable format. Officers and several hundred sailors use these devices on McFaul, proving a strong medium for workforce automation and monitoring. The wireless LAN enables workers on McFaul to exchange Microsoft Word documents, send e-mail, review technical documentation on the ship's equipment, coordinate daily and weekly schedules, and consolidate and coordinate checklists and databases. Overall, the system increases the combat readiness of the ship via the efficient management of human resources (Seybold, 2000).

Security is another application in the workplace. Sony have recently developed a Bluetooth-compliant 'wearable key' (Kyodo News, 2000). This allows people to identify themselves to computers and mobile devices so that they can access their personal data; a special wristband transmits users' identification numbers and passwords to the network system.

Applications for travel

Businesspeople typically spend a considerable amount of time in-transit. Much of this time is not used productively due to a lack of facilities to enable connectivity. However, significant research and development is now being undertaken to make vehicles 'smart'; creating connectivity with wireless voice and data capability. In the situation where the businessperson is in a car, an in-vehicle network, for example, Bluetooth, could allow connectivity of personal devices; using a mobile phone as the wide-area network connection, the employee could connect to the Internet or corporate intranet to receive e-mail or access other corporate systems (Miller, 2000). Voice recognition could automate this even further. To this end, a consortium comprising Microsoft and five Japanese companies has recently agreed to develop software for embedded wireless in the automobile industry, enabling in-car computers capable of hands-free communication, access to the Internet, and instant summoning of emergency services or roadside assistance. The consortium includes Japan's biggest car maker, Denso; Toyota-affiliated car parts maker Aisin AW; leading Japanese car audio equipment maker Clarion; Nissan-affiliated car navigation system maker Xanavi Informatics; and Japan's second biggest trading house, Mitsui (Reuters, 2000). The system for in-car networking will be available from 2002.

Within an aircraft, subway, or rail train device connectivity provides similar benefits. Plane manufacturers such as Boeing have already invested large sums in developing sophisticated in-flight entertainment and work systems (IT Week, 2001; Lui-Kwan, 2000). Here, connectivity can be provided for gaming, collaboration with fellow passengers, e-mail, and Web access. Alternatively, offline data transfer can be used. Every time the user passes near an 'infostation', such as in the airport, selected information is transferred. Goodman (2000: 39) provides an interesting example of the application of an infostation:

At my departure airport, as my laptop computer goes through the X-ray machine, the infostation downloads lots of information that might be useful to me during the flight, such as e-mail, voice mail, faxes, and reading material about the attractions and events at my destination. When my plane lands and I walk through the jetway corridor, an infostation will upload data I generated during the flight and download new material from my home server, as well as local weather and traffic reports.

Another scenario involves the baggage of an individual in transit. A practical solution for travellers from Sabre Holding and BlueTags – eBaggage Tracking – enables the location of luggage to be tracked at any stage of a journey, from start to finish. The commercial solution allows customers to electronically check their 'tagged' bags with the airline by using a special scanner, and, upon arrival after a flight, will receive messages about the whereabouts of those bags via PDA or mobile phone (10meters.com, 2001). This is a very new form of tracking, enabled by Bluetooth, replacing similar systems that use barcodes.

Commercially, there are many more applications of short-range wireless technology. Using wireless IT, a firm's inbound and outbound logistics can be accurately monitored. Wireless transceivers let Bluetooth-enabled portable terminals – such as PDAs – communicate with a central database through an in-vehicle transceiver. For example, IBM's route sales solutions allow adjusting and printing invoices from a vehicle, driver access to customer details, and the tracking of goods (a handheld system can keep track of deliveries, pickups, and returns for end-of-route reconciliation) (Farrell, 2000). Conceivably, automatic inventory scanning and calculation within a vehicle is possible, computerizing the process even further (Zeus Wireless, 2002a).

Applications in the home

Within the home, short-range wireless becomes useful for controlling a range of devices. Typically, a server wired to the Internet can be used for home security, automated meter reading, differential pricing of utilities (gas, electricity, and water), home automation with remote access, and appliance monitoring or preventative maintenance (Gaw and Marsh, 2000). These are discussed in more detail below.

Wireless transceivers, communicating with a central PC or controller from various locations, allow reliable security of indoor and outdoor facilities (Zeus Wireless, 2002a). Automatic alarms can alert services to a call-out situation. For example, a break-in could trigger an intruder alarm, instantly informing the police and alerting a nearby officer.

In the utilities industry (e.g. gas, electricity, and water), meter reading tends to be a drain on time and resources for field staff. Computerized customer phone systems for recording readings have helped to alleviate this problem to a certain extent. However, often estimates or customer readings are inaccurate, costing the company in terms of money owed. Mobile reading systems take a giant leap over this problem. This information can be used to create more advanced differential pricing structures, based on a fuller understanding of the demand for resources and cost of supply. For example, some applications for the electricity industry allow a reduced rate in exchange for allowing the utility company to turn off power to certain devices during peak periods. Typically, 'smart' thermostats could take control at peak-load times on the grid to avoid buying expensive power (Kridel, 2000).

Another area where wireless IT is likely to have an important role is in field testing and reporting (Research in Motion, 2000a). As the cost of wireless devices falls and the performance of said devices increases, impetus is provided for the inclusion of these devices in products such as cars, refrigerators, washing machines, vacuum cleaners, industrial equipment, and many other devices and appliances (US Internet Council and ITTA, 2000). Such devices will be able to store and report information on the performance of products, providing an important source for future product and technology development and refinement. An example is Ariston's margarita2000.com washing machine, which communicates using SMS. Moreover, such devices can be used to enable better service or preventative maintenance for the customer. A printer may send an alert that it is about to fail and should be replaced on the next day. An oven might call General Electric when a heating element is about to expire (Shankland, 2000). A vacuum cleaner may indicate that the model is out of date and information on a new product could be sent to the customer. If a field service engineer is sent to the customer to investigate a problem that is not fully diagnosed remotely, then the failing device might even disable the backdoor alarm to allow the repairman to enter the building as he moves into range (Shankland, 2000).

On another level, solutions such as HomeRF are also aimed at flexibility for home computer users. Wireless home networks allow multiple PCs to share Internet access, printers, files, and drives, and participate in multiplayer games without wires, offering the consumer computing flexibility and mobility. In a similar fashion to Bluetooth dongles, HomeRF devices can be plugged into PCs and peripherals to offer convenient, economical, and entertaining network computing (HomeRF, 2000).

Applications for the public

Public facilities providing nodes for wireless devices could be useful in many locations – such as shopping malls, hotels, and airports. Troy Group recently announced a partnership with InnTechnology to bring Bluetooth wireless technology to the hospitality industry. Hotel guests, a large proportion of whom tend to be business travellers or conference attendees, will be able to access in-room printers, gain LAN access, and communicate with other guests using Bluetooth-enabled notebooks, PDAs, or phones (Business Wire, 2000).

Conceivably, the electronic key concept could also be useful for access to hotel rooms.

For the roaming user, PDAs and smartphones could prove useful for mobile electronic cash. As mentioned in chapter 1, various methods exist for this including smart cards, pay-by-GSM and dual-SIM. In another trial, Ericsson and Eurocard AB in Sweden are piloting the Ericsson R520 Bluetooth phone and a virtual Eurocard for POS use (10meters.com, 2001). With the growth of local interaction between devices, buying products using a mobile phone could have the potential to become commonplace.

Other possible 'public' applications are vending and automated teller machines (ATMs) (Arthur D. Little, 2000; Research in Motion, 2000b; Zeus Wireless, 2002a,b). Typically, vending machines in a shopping mall could connect to a server hub and automatically signal to the corporate computer when they require restocking (Research in Motion, 2000b). This information could be immediately used in the supply chain to source vending products from suppliers. Restocking of vending machines, for example, in a large office building, can be managed quickly and efficiently, maximizing cash collection (Zeus Wireless, 2002a). In addition, online communications with vending machines allows other activities including cash inventory, systems status, sales information, changing product prices, and updating software (Zeus Wireless, 2002b). Similarly to vending machines, ATMs can be connected to bank networks using wireless IT, allowing seamless transactions and reporting of cash inventories (Arthur D. Little, 2000).

Customers also have the possibility of linking to ATMs or vending machines using short-range wireless technology to make transactions. One such machine is NCR's Freedom, currently being demonstrated at the Marriott Marquis in New York's Times Square (10meters.com, 2001). The machines themselves have no screen or keyboard. Consumers type their cash withdrawal requests into their PDA or mobile phone and then point the device at the ATM to get money dispensed. A security PIN number, also typed by the device itself, allows access to the account. While much of the mechanics of withdrawing cash have not changed, the process now focuses on the consumer's mobile device.

Finally, short-range wireless can be used for numerous LBS. For example, the customer looking for a specific product or price could scan or enter a code into a phone; when the customer walks past a store with the right product or price the phone could send an alert.

Further, the roaming phone user can be provided with information, alerts, or even advertisements based on their location. For example, walking down the street in an urban area could set off a plethora of messages from retailers eager to tempt clientele inside – positional (p-) commerce. Similarly, an alert could inform the user of a security threat in a certain part of the city, or information could be given on an exhibition outside an art gallery.

Opportunities and problems for short-range wireless solutions

There are a number of issues of note in the evaluation of the emerging technological solutions discussed above. This section examines some of the key benefits associated with these solutions and espoused by developers, the proposition of e-commerce via SRW, and some of the major issues in implementing solutions for the marketplace.

Benefits of short-range wireless technology

The applications examined earlier were many and various, with a variety of apparent benefits. Taken as a whole, these inter-related benefits can be summarized into eight key areas – as demonstrated in Figure 5.2.

Typically, the technologies offer a high degree of *flexibility* in the way that they are used. The small size and low-power drain of wireless units means that they can be embedded in all manner of devices to provide a new wave of adaptability in the way that people use them. Associated with this, the designs of products and solutions become more flexible. A new design ethic means that electronic components no longer need to be housed in the same or adjacent equipment. Elements that are not a key part of the user interface can be embedded and hidden surreptitiously to provide a more aesthetic or ergonomic end-product. For example, a smartphone could be manufactured as separate components: an earpiece, a screen on a wristband, and belt-buckle transceiver.

Technologies such as Bluetooth and HomeRF have also been designed with regard to *affordability*; a standard unit costs in the

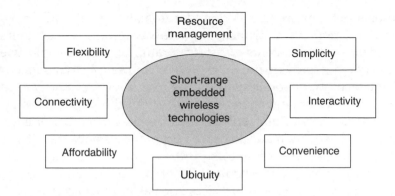

Figure 5.2   Key benefits of short-range embedded wireless solutions

region of $5 to produce. This, along with aspects such as flexibility, has a direct relationship on the ability of companies to embed the technology in a large number of products. Since *ubiquity* is a key aim of players in the Bluetooth SIG and other standards groups, this is important. In addition, the diversity of product offerings from companies in the Bluetooth SIG (e.g. mobile phones, PDAs, computers, computer hardware, and software) creates a strong platform for market penetration.

One of the main espoused benefits of wireless solutions is *simplicity*. In the current 'wired world', connectivity of electronic equipment is dictated by cumbersome and complex wiring solutions. Often wiring is proprietary, but it is always time consuming to set up electronic equipment (with risk of, e.g., static or cable damage). Removing the wires makes connection simpler; for example, Bluetooth allows automatic *connectivity* to a local network simply by bringing a device into range (such as a PDA or phone), and a large number and variety of devices are supported. Any device embedded with a suitable short-range wireless unit can link into a network and begin sharing data. Even traditionally isolated devices, such as white goods, can be connected effortlessly to provide additional services (such as information on current performance or maintenance requirements). Thus, the potential for complex *interactivity* and information sharing between devices is enormous.

As devices become more interactive and software technologies are developed to allow a fuller understanding of capabilities, this enables advanced *resource management* for devices (see information on Jini below). No longer are devices independent and they can work in partnership; for example, a PC could automatically take advantage of unused memory in a printer, a microwave could create a menu dependent on the contents of a refrigerator, or a PDA could use an audio system to provide music. Moreover, the simplicity, connectivity, and interactivity of solutions provide a high level of *convenience*. In theory, the user no longer has to think about the problems of establishing device networking, only its benefits; the business user entering a car could have automatic access to corporate systems, the office user may gain immediate access to secure systems using a wireless key, meter reading becomes unobtrusive at home, or hotel rooms can provide convenient wireless services such as Internet access.

Overall, although these are key benefits for the technologies described earlier, the solutions vary in the extent to which they fulfil them (see above). Of the four main competitors, Bluetooth and HomeRF are similar in many respects, such as price, convenience, flexibility, and simplicity. However, aside from the differences in market focus – HomeRF is essentially a home networking technology, whilst Bluetooth is more general – HomeRF is argued to have greater connectivity; typically, HomeRF can support a larger number of devices with a greater payload. Turning to the other competing wireless technologies, although supporting considerably higher payloads, IrDA and IEEE 802.11 are slightly more expensive. IrDA is also much less flexible and less capable of connectivity, requiring short-range line-of-sight. Although established with some ubiquity, IrDA also has some fragmentation. On the contrary, IEEE 802.11 provides the most superior of all connectivity in terms of the number of devices supported.

New models of electronic commerce via short-range wireless

The functional characteristics of short-range wireless outlined above feed directly into the ability of the technology to transform existing forms of e-commerce and to develop new ones. An e-commerce world without wires could look very different to the situation we know today. Figure 5.3 shows two simplified models of e-commerce: (a) before SRW and penetration of the mobile Internet, as we know it today; and (b) where SRW becomes a key mode of data exchange and mobile data communication becomes commonplace.

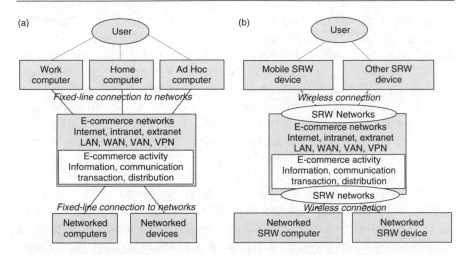

Figure 5.3 E-commerce and the user – the impact of SRW and mobile devices: (a) e-commerce prior to SRW; (b) e-commerce via SRW

In the current situation prior to widespread SRW and mobile device use, users are usually obliged to access networks through wired connections on their computers at work, at home, or though ad hoc facilities (e.g. a cyber café). The networks used to conduct electronic commerce typically include the Internet, intranets, and extranets, with possible use of, for example, local area networks (LANs), wide area networks (WANs), value-added networks (VANs), or virtual private networks (VPNs). Through the network, services may be accessed from remote networked computers and devices, such as those on the Internet, or from more local computers and devices, such as those within an organization's LAN. The basic e-commerce activities include those of sharing information, communicating, trading, and distributing products between consumers and organizations, between organizations, within organizations, and even between consumers (Barnes and Hunt, 2001).

In the situation where SRW-enabled mobile devices become pervasive a very different model of e-commerce is likely to emerge. The functional characteristics of SRW described in the previous section afford a new mode of access without wires, enabling a range of services, including those mentioned earlier in the chapter. The user with a mobile SRW device can access any other compatible devices on SRW networks within range. These networks can appear in many different locations than has traditionally been the case for ad hoc computer networking, including homes, airports, aircraft, motor vehicles, ATMs, hotels, shopping malls, and other public 'infostations'. Moreover, SRW networks can be created ad hoc by devices being brought together in a new location, for example, at a business conference. Whatever the scenario, each separate SRW network can be linked to a wired network, such as the Internet, opening the door to full e-commerce capability via the short-range device, expanding the potential for informing, communicating, trading, and product and service distribution by connection to other computers or devices, whether connected via SRW or not.

Unlike traditional e-commerce, which has tended to focus on remote services, e-commerce via SRW also enables interaction at close physical proximity. It thus has a much greater potential impact on the day-to-day physical activity of the roaming user than traditional e-commerce.

Current issues in short-range wireless application development

The above applications of short-range embedded wireless provide massive opportunities. Nevertheless, it must be remembered that these developments are still somewhat embryonic and have some way to go before they are mature enough to significantly penetrate markets. Key issues in bringing these applications to the market include the availability of suitable software, typically for service discovery, standardization, cost, network transmission requirements, and security of wireless information. In essence, these problems mirror the key benefits shown in Figure 5.2.

An important part of obtaining optimum use from wireless resources is the means for disparate devices to communicate their functional capabilities to each other. This process is referred to as 'service discovery'. The basic service discovery trait is a common way of describing the functions required by the requester and provided by other devices, applications, and services – a capability description. Typically, capability descriptions are provided for device and service classes. For example, the capabilities of a printer include paper size, print density, and colour capability. Additional characteristics include input source, output tray, and finishing functions like binding and stapling (HP Chronicle, 2000). Consistent capability description is a necessary part of achieving standardized device communication for embedded wireless.

Various companies are developing software for service discovery, but this is still at an early stage. One of the most promising solutions is Jini from Sun Microsystems. Jini introduces a new layer of device interactivity and builds on wireless networking standards such as Bluetooth. Once connected, a Jini device automatically broadcasts what it can do and how it works – a fundamental difference from the way networks operate today. Devices can then link themselves and take advantage of each other's abilities, bypassing the need for a person to inform them of all the affected components (Shankland, 2000). For example, if a printer in one part of an office is under-utilized, it could inform other printers of the situation – which, in turn, could reroute print jobs and send e-mail alerts to the owners of the jobs.

Incredibly, standardization still appears to be an issue with embedded wireless devices. This has always been a problem with IrDA manufacturers. Recently, it has also emerged for the case of Bluetooth devices (Charleston Gazette, 2000). The first Bluetooth devices to reach the consumer market – Ericsson's T36 headphones and R520 phone connector – were proprietary. Ericsson's headset can only interact with an Ericsson Bluetooth phone. Such a trend would have severe implications for the usefulness of Bluetooth, especially if larger players in telecommunications markets chose their own variants. Clearly, it will take several years and numerous hurdles before a Bluetooth device can talk to all other Bluetooth devices.

Cost is another issue; typically, early adopters of technology bear a considerable part of research and development costs. Thus, for Bluetooth, although some chips only cost around $9 (Shim, 2002), the first devices were expensive: the first IBM Bluetooth card cost $189, the first Ericsson headset (plus cellphone attachment) cost $300, and the first Motorola Bluetooth phone was around $325 (plus $190 for computer card).

The technologies outlined in Table 5.1 allow fast communication over short distances. However, many of the more advanced applications for business rely on high-speed long-distance connectivity with networks such as Bluetooth (Intercai Mondiale, 2000). In reality, these applications are currently constrained by the data rates of wired and wireless networks. Standard wired telecommunication connections are slow for data and faster connections tend to be expensive. Wireless connection is restrained by 2G networks – such as the 14.4 Kbit/s GSM used in more than half of the cellular world. Further technologies such as General Packet Radio Service (GPRS) and Enhanced Data Rates for Global Evolution (EDGE) take this rate up to 115 and 384 Kbit/s, respectively. However, the fast connections needed for bandwidth-hungry high-data applications (such those in connecting LANs and for high-quality video) will not become available until 3G transmission technologies become commonly used. Initial roll-outs of 3G technologies in many countries (such as CDMA 2000 1x) have been expensive and not particularly well subscribed.

Finally, there are some problems with information security. Researchers at Lucent's Bell Laboratories (part of the Bluetooth SIG) recently discovered security loopholes in Bluetooth devices. Specifically, it would be possible to bug a device so that an intruder could easily obtain its encryption key. Since many applications depend on secure information this is an important issue to be resolved. Future developments are aimed at enhancing the Bluetooth technical standard so that a devices' 'identity numbers' are masked and become constantly changing pseudonyms when transmitted (Batista, 2000).

Summary and conclusions

A new wave of device interactivity is on the horizon. Whilst some of the ideas for device interactivity are not new – such as IrDA – others represent a significant departure from the current 'wired world'. By combining ideas of interactivity from computer networking and wireless telecommunications, a variety of cheap, flexible, and low-power standards are becoming available for everyday use. This chapter has examined the available standards, including Bluetooth, IEEE 802.11, IrDA, DECT, HomeRF, and UWB, attempting to summarize these technologies into a comprehensible format. Moreover, such standards have a variety of advantages and disadvantages that have been touched upon.

Building on this knowledge, the chapter has also examined the current state-of-the-art in the application of said technologies. These are many and varied, including applications in the wireless workplace, for home networking, for in-transit vehicle connectivity, and for management of publicly accessible services. Wireless workplaces offer the potential for tremendous flexibility unrestrained by the current generation of wired devices, which will be particularly useful for ad hoc group-work. Home networking allows considerable control of all manner of electronic equipment within the home or remotely to add service – for example, to control heating, for security, or for entertainment (such as remotely recording a television programme). Vehicles enabled with sophisticated computer networking facilitate an extension of the businessperson's office, allowing increased productivity for the employee in-transit. Even for public services short-range wireless can contribute in managing equipment in public places, providing public information or enabling m-commerce.

These technologies are still in the early stages of development, but they provide some compelling benefits: flexibility, connectivity, interactivity, affordability, simplicity, convenience, ubiquity, and advanced resource management. Nevertheless, at least to begin with, there are still important hurdles to achieving some of these benefits – such as those of cost, security, and standardization. These issues need to be solved before devices begin to penetrate consumer markets. In addition, advanced functionality of devices depends on sophisticated software standards and high data rates for interactivity. However, the technologies that permit this – such as Jini and UMTS – have yet to be offered fully in the commercial market.

Clearly, it will be years before technologies such as Bluetooth begin to fully deliver the promises of applications discussed earlier. It will take time for the technologies and markets to mature enough for advanced applications (Veitch, 2001). Nevertheless, industry analysts predict that this could happen sooner than may be expected. Revenues from Bluetooth products are predicted to rise from $2 billion in 2001 to $333 billion in 2006, as shipments rise from 4.2 million to 1.01 billion (Frost and Sullivan, 2001). When markets and technologies do mature, we are likely to encounter even more creative and interesting commercial applications.

The next few years will be important in determining which technologies or applications become pervasive and dominant. Whatever the case, it is clear that embedded wireless has an important part to play in the IT strategies of many organizations in the future; a number of the applications presented in this chapter are being successfully used or piloted. During 2002, WiFi was fast becoming the preferred method of data connectivity for enterprise wireless applications and beginning to move into the consumer market. The predominant value proposition is that of connectivity via 'hot-spots' in the office, car, café, and so on.

Acknowledgements

An earlier version of this chapter appeared as: Barnes, S.J. (2002). Under the skin: short-range embedded wireless technology. *International Journal of Information Management*, **22**, 165–179.

References

Anoto (2000). *The Anoto White Paper*. Lund: Anoto AB.

AnyWhereYouGo (2000). Bluetooth – frequently asked questions. http://www.anywhereyougo.com/, accessed 15 October 2000.

AU System (2000). Bluetooth White Paper 1.1. http://www.ausystem.com/, accessed 15 January 2000.

Arthur D. Little (2000). *Serving the Mobile Customer*. London: Arthur D. Little.

Barnes, S. and Hunt, B. (2001). Preface: putting the e- in business. In S. Barnes and B. Hunt, eds, *E-Commerce and V-Business: Business Models for Global Success*. Oxford: Butterworth-Heinemann, pp. ix–xvii.

Batista, E. (2000). PDA: 'public' display assistant? *Wired News*. http://www.wirednews.com/, accessed 11 September 2000.

Business Wire (2000). Technology will allow cable-free connectivity to in-room printers, copiers and fax machines. http://www.bluetooth.com/, accessed 22 June 2000.

Charleston Gazette (2000). The Bluetooths are coming. http://www.anywhereyougo.com/ayg/ayg/wireless/Article.po?id=699857, accessed 28 October 2000.

Durlacher Research (1999). *Mobile Commerce Report*. London: Durlacher Research.

Farrell, A.J. (2000). Mobile computing in the route sales marketplace. *IBM Pervasive Computing White Paper Series*. http://www-3.ibm.com/pvc/tech/mobile_computing.shtml, accessed 23 June 2000.

Frost and Sullivan (2001). End-user perceptions of Bluetooth. http://www1.frost.com/prod/corpnews.nsf/0/EACEE6D2F27FB9B280256A780049E3BC, accessed 15 July 2001.

Gaw, D. and Marsh, A. (2000). *Low-Cost Multi-Service Home Gateway Creates New Business Opportunities.* Sausalito, CA: Coactive Networks.

Goodman, D.J. (2000). The wireless Internet: promises and challenges. *IEEE Computer*, **33**, 36–41.

HomeRF (2000). HomeRF: frequently asked questions. http://www.homerf.org/wireless_faq.pdf, accessed 15 August 2000.

HP Chronicle (2000). Dynamic networking requires comprehensive service discovery. http://www.anywhereyougo.com/ayg/ayg/wireless/Article.po?id=699737, accessed 28 October 2000.

IBM Pervasive Computing (2000). Extending SAP systems to pervasive computing devices. *IBM Pervasive Computing White Paper Series.* http://www-3.ibm.com/pvc/tech/sap.shtml, accessed 28 October 2000.

IDC Research (2000). Global PC market shows steady growth. http://www.idc.com/Hardware/press/PR/PS/GPS102300pr.stm, accessed 23 October 2000.

Intercai Mondiale (2000). *Bluetooth as a 3G Enabler.* Marlow: Intercai Mondiale Ltd.

IT Week (2001). Air travellers get wireless access. http://news.zdnet.co.uk/story/0,,t294-s2091377,00.html, accessed 17 July 2001.

Kridel, T. (2000). Telemetry gets smart. *Wireless Review.* http://www1.telecomclick.com/magazineclick.asp?released=18258magazinearticle=3236, accessed 1 February 2000.

Kyodo News (2000). Sony develops wearable key. http://www.anywhereyougo.com/ayg/ayg/wireless/Article.po?=666836, accessed 26 October 2000.

Lambert, M. (2000). *Positioning Bluetooth Amongst Other Current Wireless Technologies.* Cambridge: Cambridge Consultants/Arthur D. Little.

Lui-Kwan, G. (2000). In-flight entertainment: the sky's the limit. *IEEE Computer*, **33**, 98–101.

10meters.com (2001). Bluetooth products that could make cents. http://www.10meters.com/bluetooth_bytes.html, accessed 30 September 2001.

Merritt, R. (2002). 802.11b spike reported as combo WLAN chips near. http://www.commsdesign.com/story/OEG20021024S0037, accessed 20 December 2002.

Miller, B.A. (2000). Bluetooth applications in pervasive computing. *IBM Pervasive Computing White Paper Series.*

http://www-3.ibm.com/pvc/tech/bluetoothpvc.shtml, accessed 1 February 2000.

Newsbytes News Network (2000). First Bluetooth-compliant print server arrives. http://www.anywhereyougo.com/ayg/ayg/wireless/ Article.po?id=628720, accessed 25 October 2000.

Research in Motion (2000a). Building a business case for wireless. http://www.rim.net/, accessed 15 August 2000.

Research in Motion (2000b). The wireless workforce. http://www.rim.net/, accessed 15 August 2000.

Reuters (2000). Microsoft and the Japanese rev up Windows CE for cars. http://news.cnet.com/news/0-1006-200-2813431.html, accessed 19 September 2000.

Roper Starch (2000). Global PC ownership rises but digital divide remains. http://www.roper.com/news/content/news204.htm, accessed 30 September 2000.

Seybold, A. (2000). Mobile implementation. *Outlook*, August, 1–5.

Shankland, S. (2000). Jini's bottleneck. http://news.cnet.com/news/0-1003-201-1559726.html, accessed 15 March 2000.

Shim, R. (2002). Bluetooth to break through gum line. http://news.com.com/2100-1040-942906.html?tag=rn, accessed 20 December 2002.

US Internet Council and International Technology and Trade Associates (ITTA) (2000). *State of the Internet 2000*. Washington, DC: ITTA Inc.

Veitch, M. (2001). Bluetooth five years from being ubiquitous. http://news.zdnet.co.uk/story/0,,t294-s2092159,00.html, accessed 20 September 2001.

Zeus Wireless (2002a). *Wireless Data Telemetry*. Maryland: Zeus Wireless Inc.

Zeus Wireless (2002b). Vending. http://www.zeuswireless.com/html/apps/vending/vending.html, accessed 6 October 2002.

Chapter 6

Location positioning technologies and services

Introduction

In the emerging m-commerce economy, the knowledge of the position of a given service subscriber making a call is gaining particular interest among mobile operators who can, in turn, provide innovative LBS, typically with the assistance of third parties such as service or content providers (see Barnes, 2002). Such ideas are not new. Location (l-) commerce has existed in a limited form for more than 20 years. The pioneers of LBS were basic tracking services and Automated Vehicle Location (AVL). In 2000, more than 100 companies were providing AVL products and services in the United States alone (Airbiquity, 2000a). However, until recently, the specialized location-based industry survived as a niche market to both high-end businesses (such as trucking and freight) and well-to-do customers (via automobiles such as Lexus and BMW). Typically, high-priced devices required subscriptions to special location services, suppressing demand (Frost and Sullivan, 2003).

Large-scale commercialization of location-aware services has only been recognized in the early twenty-first century, as a series of events and trends have begun to provide an environment that is conducive. Underlying the growth in commercial LBS markets are recent technological advancements (in handsets, networks, and positioning technologies), regulatory change (including derestriction of satellite positioning technologies and mandates for emergency services), industry trends (particularly the need for new value-added services, mergers/acquisitions, and call-centre development), and emerging business opportunities (as a result of converging market conditions, e.g. the growth of LBS in Japan driven by the popular iMode service). As a result, the door has been opened to a vast array of commercial applications, including those for emergency services, asset tracking, navigation, location-sensitive billing, and location-based information services. Indeed, the Strategy Analytics (2003) estimates that LBS could be worth $8 billion by 2008.

Recently, the terrorist attack of 11 September 2001 has highlighted the value of LBS technology. In response to the emergency, cellular networks were heavily used both by the public and rescue workers. Of particular note were the jury-rigged automatic location systems used in New York as part of the rescue effort. Undoubtedly, the 11 September 2001 attacks have helped focus attention on the importance of a system that locates emergency callers.

The purpose of this chapter is to examine the emerging l-commerce phenomenon. To this end, it analyses the technologies and applications involved with introducing the new wave of LBS (provided in the next two sections). The chapter continues by exploring a value proposition model for services and some of the core inhibitors. It also briefly explores some of the strategic business implications of these services. Finally, the chapter rounds off with some conclusions and predictions regarding the future of l-commerce.

Location technologies for mobile commerce

One or more location methods can be used to determine the position of user equipment for LBS. It is also possible to distinguish between methods that are most useful inside and outside buildings. Leading candidates for indoor location identification include short-range radio, such as Bluetooth, and Infrared (IR) sensors (Barnes, 2002). For example, developers could use Bluetooth or IR to build an automatic tour-guide system, such as for an art gallery; as the tourist with a suitably enabled PDA device moves into range of a piece of artwork, it could send out a signal that automatically displays information related to the artwork on the screen (Tseng *et al.*, 2001). However, interesting though this is, the focus of this chapter is on roaming, location-aware technology used largely outside buildings.

For a detailed examination of the benefits and applications of short-range wireless technologies see chapter 5.

Location techniques operate in two steps – signal measurements and location estimate computation based on the measurements – which may be carried out by the user equipment or the telecommunications network (Mitchell and Whitmore, 2003). Subsequently, positioning techniques can be categorized into several varieties, each with its advantages and disadvantages. The main types are cell-location, advanced network-based, and satellite-based positioning. Three of the main categories of positioning methods are shown in Table 6.1, in order of increasing accuracy.

Cell-location positioning techniques

This technique works by identification of the cell of the network in which the handset is operating (the 'cell of origin'). Cell of origin (COO), sometimes called Cell-id, is the main technology that is widely deployed in wireless networks today. It requires no modification to handsets or networks since it uses the mobile network base station as the location of the caller. However, although locating the caller is fast – typically around 3 seconds – accuracy is limited. Positioning accuracy depends on the size of the cell and techniques used for enhancing location calculation, such as user self-locating (where by end-users use landmarks and addresses to improve their positioning precision) and propagation time measurements. Position accuracy down to 150 m in urban areas is not uncommon, growing very significantly outside major areas of population.

Advanced network-based positioning techniques

Advanced network-based techniques rely on the measurement of signals from nearby base stations via the user's equipment. The position of the user is derived by triangulation, using techniques such as Enhanced Observed Time Difference (E-OTD) and Observed Time Difference of Arrival (OTDOA). The E-OTD method works on the GSM network. Variations of E-OTD such as Advanced Forward Link Triangulation (AF-LT) and Idle Period Downlink (IP-DL) have been developed for CDMA and WCDMA networks. The positional

Table 6.1 Three classes of location positioning

Location service (LS) category	Explanation	Typical methods of positioning	Accuracy	Response time	Key limitations	Market requirements
Category LS1 (basic service level)	Location of all handsets with at least cell accuracy	Cell of Origin (COO) or Cell-id, including Service Area Identity (SAI), LocWAP, and enhanced Cell-id. May also include enhancements with propagation time measurements	Low. Depends on cell size and enhancements; typically 150–10,000 m	Very fast. Typically around 3 seconds	Very limited accuracy in areas with low cell radius	No modifications needed to networks or handsets
Category LS2 (enhanced service level)	Location of all new handsets with reasonable cost and improved accuracy	Estimated Time of Arrival (EOTD) for GSM, and its variations such as Advanced Forward Link Triangulation (AF-LT) and Idle Period Downlink (IP-DL) for CDMA and WCDMA, respectively	Medium. Typically around 50–125 m	Fast. EOTD takes around 5 seconds	Dependent on visibility of base stations for signal measurement and number of location measuring units (LMUs)	Software-modified handsets needed for positioning
Category LS3 (extended service level)	Location of new handsets with high accuracy and higher costs than LS2	Global Positioning System (GPS) and Assisted Global Positioning System (AGPS)	High. Outside buildings, approx. 10–20 m; inside buildings, approx. 50 m	Variable. GPS takes around 10–60 seconds, but AGPS around 5 seconds	Signal degradation and reduced accuracy in certain environments, e.g. inside buildings or 'urban canyons'	New handsets needed for positioning

information is based on relative times of arrival of signals at the handset and fixed receivers as sent by base stations. Location receivers or reference beacons (referred to as Location Measuring Units or LMUs) are overlaid on the cellular network at a number of geographically dispersed sites. Location is then calculated using the time differences of arrival of the signal from each base station at the specially enabled handset and LMU (via time stamps and intersecting hyperbolic lines). E-OTD is typically accurate to approximately 50–125 m, with a response time of around 5 seconds. In a manner similar to E-OTD, OTDOA location works by calculating the time difference of the arrival of a signal from a mobile device and three network base stations. The large cost of network synchronization affords only small improvements over COO in urban areas, and the response time is much higher at around 10 seconds.

Satellite-based positioning techniques

In some cases, a global navigation satellite system such as the Global Positioning System (GPS) can be used to enhance the accuracy of radio positioning. GPS has been available for general use since the early 1990s. Operating in the L-band frequencies GPS can be used anywhere in the world. The system's satellites transmit navigation messages that contain their orbital elements, clocks, and statuses, which a GPS receiver uses to determine its position and thus its roaming velocity (Tseng *et al.*, 2001). Determining the receiver's longitude and latitude requires three satellites, and adding a fourth can determine the user's altitude. However, only recently (in May 2000) has the US Army derestricted outdoor positioning to a sufficiently high resolution for advanced use – currently 10–20 m.

Stand-alone GPS has the key problems of no indoor coverage and a relatively long time to first fix, usually 10–60 seconds. It also fails in radio shadows and requires considerable cost, complexity and battery consumption in handsets (Djuknic and Richton, 2001).

The same issues are also involved in the use of GPS for mobile ad hoc networks (MANETs). A MANET consists of a set of mobile hosts that roam at will and communicate with one another. Typical examples of MANET applications include battlefields, festival grounds, assemblies, outdoor activities, rescue actions, and major disaster areas, where communications are needed immediately without core network infrastructure (Tseng *et al.*, 2001). However, the flexibility of these systems could lead towards more developed forms being further used commercially. Communication in MANETs takes place through wireless links among mobile hosts, using their antennae, but no base stations. Transmission limitation means that several hosts may be needed to relay a packet between sender and receiver (Tseng *et al.*, 2001). In this environment, location-positioning technologies are needed that do not require traditional network infrastructure. GPS is the prime enabler for this type of outdoor positioning.

Using GPS in addition to a wireless network – often referred to as assisted-GPS (AGPS) – can provide significant extra benefits. Embedding a GPS receiver into the user's handset can directly provide positioning fixes in less than 5 seconds; the network may assist the user equipment by reducing the power consumption of the handset, by optimizing the start-up and acquisition time and by increasing the sensitivity of the GPS device (Lavroff, 2000). AGPS can also be used indoors, where it is accurate to within 50 m. In the future, technologies such as Bluetooth and IEEE 802.11 may enable assisted location positioning within buildings and hot-spots to even higher resolutions, suggested at around 10 m (Nokia, 2003).

Overall, the various positioning technologies are complementary – there is no single universal solution. Where both accuracy and coverage are important, hybrid technologies may provide an optimum solution. Cellular and advanced network-based technologies can be used to fill in the gaps in coverage from satellite-based systems, like GPS. The basic positioning accuracy category is focused on market penetration, and should be available for all phones, enabling a fast time-to-market. The intermediate category will have a software impact on handsets, whilst the high-accuracy category will have a hardware impact on handsets. All three levels of accuracy will exist in parallel in the future (Nokia, 2003).

Applications of l-commerce

The kinds of location-based technologies described above enable many advanced forms of data services based on the position of the user, in both personal and business markets. Typically, services can be categorized into four main areas, as shown in Table 6.2. Let us examine each of these areas in turn.

Table 6.2 Key areas of application of location-based services

Area of use	Application	Purpose
Safety	Emergency services	Obtain help from emergency services
	Roadside assistance	Obtain breakdown assistance
Navigation and tracking	Vehicle navigation	Reach a destination
	Fleet management	Manage fleet resources
	Asset tracking	Locate and direct assets
	People tracking	Locate and direct people
Transactions	Location-sensitive billing	Competitive pricing
	Zone-based traffic calming	Automatic pricing of road usage
	Cross-selling	Sales of products and services
Information	Locational advertising	Targeted advertisement
	Public infostation	Provide public information
	Geographic messaging	Localized information and alerts
	Yellow pages	Find proximity of something specific

Safety

The prime driver for the implementation of l-commerce infrastructure in the United States is safety. Emergency and rescue services have a vital need to know the current location of any host that sends an emergency message. The US Government has mandated that providers of personal communication systems must, in the near future, add location-identification capability to their emergency 911 (E911) services (Rockwell, 2003). Specifically, the original mandate stated that handset-based solutions must, by 1 October 2001, locate an emergency caller to within 50 m for 67 per cent of calls and within 150 m for 95 per cent of calls. Alternatively, the carriers relying upon network-based technology must achieve location accuracy of 100 m for 67 per cent of calls and 300 m for 95 per cent of calls. Location platforms such as Xypoint, the largest provider of such services to operators, are pioneers in this area (Xypoint, 2001a,b). Carriers must also undertake reasonable efforts to achieve 100 per cent penetration of handsets allowing location services by 31 December 2004 (Lavroff, 2000). Europe has no such mandate, but the European Commission is considering a less vigorous version of E911 as part of its regulatory telecommunications framework and enhanced 112 initiative (WLIA, 2003a).

The same technology used for E911 services also has value for other, related aspects of personal safety, particularly roadside assistance. In the event of an emergency breakdown or accident, the consumer's mobile device could be used to assist in getting roadside assistance to the right location. Accuracy requirements are likely to be around 125 m or less for mass acceptance, although 500 m is widely regarded as the entry-level requirement for such services.

Navigation and tracking

Driving directions and the tracking of fleet, packages, and people are a core segment of the emerging LBS market. In the United States, the average person spends 500 hours a year in an automobile. Interestingly, though, only 100,000 of the 146 million registered cars in the United States and 20 per cent of fleet trucks are equipped for telematics (i.e. LBS for automobiles). Key players include ATX Technologies/Protection One, AAA/Response, Signature, Cross Country, and OnStar commercial call centres (Airbiquity, 2000a). With the widespread deployment of services, the telematics market could expand considerably in the next few years.

Location technologies can play an important part in logistics. Intelligent transportation systems are being introduced around the world, and location technology plays a key part in almost every solution. Taxis are being equipped with automatic vehicle-location devices, allowing the fleet dispatch system to automatically select the taxi closest to the pickup location (Research in Motion, 2000). Similarly, fleet management systems are helping freight companies to monitor the status of deliveries and other logistics activities (Varshney and Vetter, 2002). Wireless transceivers let portable terminals – such as PDAs – communicate with a central database. Terminals can log in shipments of materials from vendors and track those materials in inventory, as they are needed (Zeus Wireless, 2000). In this situation, it is even conceivable to know all inventory in transit – or 'rolling' inventory – allowing an efficient method of selecting a source of components based on their known location. By knowing the location of 'rolling' inventory, times between transaction, manufacture and delivery can be further reduced (Varshney, 2000).

In Israel, mobile operator Orange offers a wireless workforce application, using CT Motion's Cellebrity platform, which includes location technology based on EOTD. Orange can offer companies the ability to monitor the movements of their workforce throughout the country to an accuracy of up to 100 m. The central coordinator, or dispatcher, can see where each of the workers are on a map of the country and so can allocate tasks more efficiently (Wieland, 2000).

Tourists are a key customer segment requiring location-based information, since they are most often found in unfamiliar geographic areas. Some services, such as Bluesigns, are aimed at these consumers (http://www.tcial.com/bluesigns/). Bluesigns works by the tourist phoning the Bluesigns tourist information centre via a telephone access number. During the tourist's call, location is determined and location-sensitive information is generated from a database. Location can be determined either by GPS or verbal communication with an operator at the tourist information centre. Based on the customer's location, the tourist can be guided along highways to a particular destination, such as a petrol station, restaurant, hotel, or tourist attraction.

Similarly, Webraska – a mapping and navigation company – offers a number of services through mobile operators such as KPN (Holland), AirTel (Spain), SFR (France), and Proximus (Belgium). Based on COO technology, the service requires users to type in abbreviations of their location (e.g. L-O-N for London, then B-A-K for Baker Street) before the network can locate roughly where they are and provide location-specific information such as directions (Wieland, 2000).

Transactions

Location-based transactions are perhaps the most complex set of services. The main thrust of the business model is billing based on the customer's location. For example, a number of countries, such as Singapore, use road pricing as part of their traffic calming and environmental policy. Payments are made electronically through a special, in-car device on entry into a particular geographic area requiring payment. If secure payment mechanisms become generally available through the mobile phone, e.g., secure electronic cash, this could present a convenient replacement for these proprietary, traditional systems.

Location-sensitive transactions opens the way to new forms of price differentiation based on the location of the user. On one level, BellSouth Wireless Data offer 'distressed items' such as dis-counted last-minute tickets to Broadway shows for people who are near enough to pick them up before curtain (Bourrie, 2000) – a form of price discrimination. On another level, individuals could be charged and taxed according to geographic region, such as a US state or country, or proximity to outlets selling goods that the con-sumer wishes to purchase.

Location-based cross-selling is another possible stream of transac-tion revenue. For example, the mobile user who has just seen a film at the cinema could immediately be offered a compact disc (CD) or digital video disc (DVD) of the soundtrack or film. Similarly, in addition to charging for information requests, such as a query for a restaurant address, service providers could earn additional revenue by asking subscribers whether, for another five cents, they would like directions to the restaurant. Ultimately, retailers, such as restaurants, could share in service costs to encourage customer interest.

In terms of payment systems, the m-Internet has some way to go towards maturity. However, as we have seen in chapter 1, a number of solutions are underway. A couple of novel payment systems are those of Paybox AG from Germany (www.paybox.net) and Mint AB from Sweden (www.mint.nu). While the latter is localised, Paybox operates in Germany, Austria, Spain, Sweden and the United Kingdom. Both are independent commercial services. Paybox uses an integrated voice recognition (IVR) system. A user making a trans-action provides her mobile number to a merchant, who then calls the IVR with the transaction data. The user receives an IVR call asking for a PIN code. During 2002, Paybox had more than 750,000 sub-scribed users and 10,000 available access points.

Information

The roaming user can be provided with information, alerts, or even advertisements based on their locale. Typically, advertisements depend on location. For example, a particular sale may interest only people within a certain distance of a merchant's store. Thus, the sender will only need to transmit the advertisement – which can be regarded as a broadcast message – to users within a set distance (Tseng *et al.*, 2001). For example, walking down the street in an urban area could set off a plethora of messages from retailers eager to tempt clientele inside. NTT DoCoMo, among others, is currently piloting such ideas via digital couponing, offering discounted products and services to subscribers within a certain radius of participating merchants. Similarly, GeePS is beta-testing its location-based wireless online shopping portal in New York and San Francisco, using couponing and other strategies.

Similar to advertising, geographic messaging is another useful application of location technologies. For example, an alert could inform the user of a security threat in a certain part of the city such as a train station, stadium, or shopping mall. Other types of public localized information can also be broadcast in a particular area – a public infostation; for example, the opening times of a public library, movie theatre listings, city phone directories, the schedule of bus services, or the availability of parking spaces could all be public broadcast information.

One of the most basic LBS offered by mobile operators is the mobile Yellow Pages. Indeed, many European operators are reluctant or unable to go beyond this sort of service, offered, for example, by Sonera, diAx, and Telia. In this type of service, the roaming user asks the question: 'what's near me?' For example, items such as locations of restaurants, shops, public transport, or nearby ATMs may be useful to the user as they move through an unfamiliar city. Weather or traffic information can also prove useful; Bell Mobility's Book4golf service allows the user to locate a North American golf course, book a tee time, and get a location-specific weather forecast.

An extension of the 'what's around' type of service is the 'who's around' service. Such an application determines who currently occupies a specific geographic area. These services are useful for planned or unplanned rendezvous between individuals, such as business colleagues or social friends. Meeting (or possibly avoiding) people becomes much simpler if individuals are enabled for such LBS. Most recently this has been used for mobile dating services.

Creating relevant user services – a value proposition model

While geo-location technologies open the door to a variety of services in consumer and business markets, location in isolation provides a bounded set of opportunities, and the potential for developing relevant services for the user goes much farther in its scope. Indeed, some of the LBS discussed in the previous section have hinted at some other important value propositions for wireless services. Overall, wireless devices have a number of features that together form a fertile bed for advanced value-added services (Kannan *et al.*, 2001). Typically, devices are very personal to the user, and carried on the person; aspects of the context of the user, such as time and place, can be measured and interpreted; services can be provided at the point of need; and applications can be highly interactive, portable, and engaging. For the consumer, this means that wireless services can be potentially very personal, timely, and relevant, or even integrated with other services in a near-seamless way (Katz-Stone, 2001).

In terms of the value proposition, we can untangle three important aspects that influence the nature of service relevance: time, location, and personal characteristics of the user. An individual's behaviour is likely to be influenced by their location, time of day, day of week, week of year, and so on. Individuals may have a routine that takes them to certain places at certain times, which may be pertinent for mobile services. If so, marketers can pinpoint location and attempt to provide content at the right time and point of need, which may, for example, influence impulse purchases (Kannan *et al.*, 2001). Feedback at the point of usage or purchase is also likely to be valuable in building a picture of time-space consumer behaviour. Further, the nature of the user, in terms of a plethora of personal characteristics such as age, education, socio-economic group, cultural background, residence, memberships, and so on is likely to be an important influence on how information is processed. Some of these aspects have already proven to be important influences on Internet use (OECD, 2001), and, as indicative evidence has shown, elements such as user age are proving an important influence on data communications via mobile devices (Funk, 2000; Puca, 2001). The wireless medium has a number of useful means for building customer relationships. Ubiquitous interactivity can give the customer ever more control over what they see, read, and hear. Personalization of content is possible by tracking personal identity and capturing customer data; the ultimate goal is for the user to feel understood and simulating a one-to-one personal relationship.

Let us examine a simple example, where the user wants to catch the next train from work to home. Figure 6.1 presents the value proposition model and analyses the value (*V*) attributed by the user

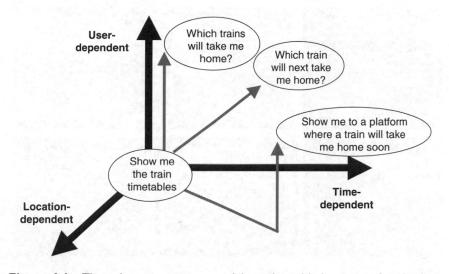

Figure 6.1 The value proposition model – value-added services for the train commuter

to a number of service options. The lowest value proposition involves the traditional provision of train schedules to the user, devoid of context ($V = a$). This is the situation at the origin of Figure 6.1. By considering further value functions (f), we may be able to create additional value associated with the service options. Adding personal characteristics (P) to the situation would enable a higher value proposition, allowing the user to ask for trains that pass the individual's home station ($V = a + f(P)$). Adding time criticality (T) to the user's options can generate an even higher value proposition ($V = a + f(P, T)$). Here, the user could, for example, ask for trains travelling home soon. Finally, adding location-dependence (L) to the user's request can create the highest value proposition ($V = a + f(P, T, L)$). Thus, the individual could ask for directions to a platform where they could catch the next train home.

Various similar value propositions are available in services on the m-Internet. Kizoom, the online travel service, provides a good example. The WAP service allows customized, time- and location-sensitive planning of travel. Figure 6.2 gives an example of how a user of the site might find details of the next train from his or her workplace to home, based on a known personal profile. The user may also go on to buy a train ticket. Essentially, the site relies on content provided from national timetables, journey planners, location services, spatial database maps, real-time travel information feeders,

Figure 6.2 Catching the next train home using the Kizoom mobile site

personal alert services, advertisers, transport companies (involved in m-commerce ticket sales), operators and m-commerce portals (Kizoom, 2000).

Other mobile services provide value propositions varying in the value-added functions of time, space, and personal characteristics. Figure 6.3 provides a simple categorization of services using the value proposition model. Focusing on LBS, there are a variety of services that are largely based around the value proposition of location-dependence. Such services include mobile yellow pages, for example, restaurant guides or nearest ATM, and location-based messaging, for example, security alerts or broadcast advertising. Adding time dependence into the value function broadens the scope to include more sophisticated services such as emergency E911, real-time traffic information, roadside assistance, and parking information. Finally, personalization opens the door to further bespoke services. The opportunities here are very large and include targeted advertisement, personal navigation aids, personal scheduling, and many other personalized services and applications. It is important to bear in mind that other services exist outside of LBS that may capitalize on aspects of the value proposition model. For example, news is largely time-dependent, mobile banking is dependent on the user and account details, and trading on a stock portfolio is dependent on both of these aspects.

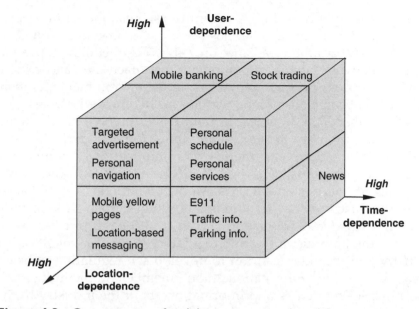

Figure 6.3 Categorization of mobile services using the value proposition model

Issues in the commercialization of LBS – privacy and standards

Although considerable progress is being made in the commercialization of LBS, they are still very much at an embryonic stage of development. Further advancement of l-commerce requires overcoming a significant number of obstacles in technology, markets, and policy. Even before companies begin to examine whether customers are willing to pay for these new services, they need to establish a technological and legal and ethical platform for service provision (O'Connor and Godar, 2003). Key areas of discussion among industry players are privacy and technology standards.

The location industry is currently in the ridiculous and destructive situation that every location finding system and positioning vendor has a different proprietary location finding technology. Recently, a number of companies have come together to form industry associations aimed at establishing standards and discussing other important industry-wide issues. In September 2000, Nokia, Ericsson, and Motorola announced the formation of the Location

Interoperability Forum (LIF), a forum to establish global interoperability standards for mobile positioning systems and solutions (http://www.openmobilealliance.org/lif/). LIF members represent a mix of network operators, equipment manufacturers, and service providers responsible for deploying equipment. Prominent members include Cambridge Positioning Systems, CellPoint, and Airflash.

In December 2000, eight leading companies involved in the wireless location industry in the United States, Canada, and Europe – Cell-Loc, SignalSoft, GoAmerica Communications, Cambridge Positioning Systems, Zero Knowledge Systems, Indexonly Technologies, iProx, and ViaVis Mobile Solutions – established another industry group, the Wireless Location Industry Association (WLIA) (http://www.wliaonline.com/). WLIA interfaces with government, administrative, and regulatory bodies on behalf of the industry and provides a forum to develop self-regulating policies, network, and share information among members in the industry. It also provides references and information about the industry to the public and policymakers, both in the United States and elsewhere.

High on the agenda for the WLIA is the issue of privacy, which promises to be a considerable challenge on wireless devices (WLIA, 2003b). In the United States, Fair Location Information Practices (FLIP) dictate that companies must: (a) inform customers about collection practices; (b) give the customer choice regarding any uses of the information; (c) allow for access to the data so that customers can ensure that it is correct; (d) maintain the data securely; and (e) comply with enforcement and auditing of the FLIP policies (Airbiquity, 2000b). Recent attempts to introduce further wireless privacy policies, such as the Wireless Telephone Spam Protection Act (aimed particularly at unsolicited wireless advertising), could further limit the use of location-based technology for marketing to mobile users (Greening, 2003). Overcoming such significant hurdles in the United States in order to reach the commercial market will prove difficult. In Europe, no comparable policies exist yet, but they are currently under discussion at the European Commission.

Strategic implications of LBS

Ultimately, the implementation of LBS technologies and applications has significant implications at the strategic business level. One strategic issue is the selection of an appropriate technology for a particular service, both in terms of location positioning infrastructure and speed of network. Most LBS, except those for safety and navigation, can begin with cell-level accuracy and the current second generation of networks. However, for mass acceptance, the technological platform for specific LBS must go beyond this. To this end, Table 6.3 shows a variety of LBS applications and indicates some of the appropriate platforms. Typically, packet-switched networks – such as GPRS and CDMA 2000 1x – are more suitable for applications where short, intermittent bursts of data are required, such as navigation, tracking, or intermittent messaging. Other applications can use any network, although the requirements for network speed in areas such mobile Yellow Pages and cross-selling will rise with consumer demand for multimedia. Accuracy depends on the criticality of location in an application. This is highest where the exact location of an individual's handset needs to be known, for example, emergency services or navigation. More general requirements for zonal targeting reduce the required accuracy.

Once a technological platform is in place that supports the strategic objectives of the firm, the benefits that accrue in developing and using LBS can occur in many parts of the organization. For example, existing services can become more efficient (e.g. emergency calls), processes can be transformed (e.g. logistics), and new services can

Table 6.3 Typical technology requirements for location services

Application	Typical accuracy requirement (typical technology)	Typical network type (typical technology)
Emergency services	High (AGPS)	Any
Roadside assistance	Medium (EOTD)	Any
Vehicle navigation	High (AGPS)	Packet (GPRS+)
Fleet management	High/medium (AGPS/EOTD)	Packet (GPRS+)
Asset tracking, e.g. packages	Low (COO)	Packet (GPRS+)
People tracking, e.g. workers	Medium/low (EOTD/COO)	Packet (GPRS+)
Location-based advertising	Medium/low (EOTD/COO)	Packet (GPRS+)
Public infostation	Medium/low (EOTD/COO)	Packet (GPRS+)
Geographic messaging	Medium/low (EOTD/COO)	Packet (GPRS+)
Yellow Pages	Medium/low (EOTD/COO)	Any
Location-sensitive billing	Medium/low (EOTD/COO)	Any
Road pricing	Medium (EOTD)	Any
Cross-selling	High/medium (AGPS/EOTD)	Any

Area of impact

	Individual/group	Function	Organization
Efficiency	*Task mechanization,* e.g. emergency call	*Process automation,* e.g. package tracking	*Boundary extension,* e.g. fleet management
Effectiveness	*Work improvement* e.g. navigation	*Functional enhancement* e.g. rolling inventory	*Service enhancement,* e.g. infostation, messaging
Transformation	*Role expansion,* e.g. workforce coordination	*Functional redefinition,* e.g. supply-chain redesign	*Product innovation,* e.g. location-based products

Benefits

Figure 6.4 Benefits of LBS applications – Index Matrix

be developed (e.g. location-based products). Figure 6.4 analyses the strategic benefits from LBS using the Index Matrix, which categorizes the benefits in terms of efficiency, effectiveness, and transformation, in the areas of the individual, function, and organization (Farbey *et al.*, 1992). Note that, where an LBS application could be placed into several cells, it has been placed into the cell of best fit for illustration. As the matrix demonstrates, the strategic impact of LBS is very deep, and applications can provide significant strategic benefits ranging from basic efficiency improvement to complex redefinition and redesign of organizational aspects.

Large transformation benefits are proving difficult to achieve, requiring a significant amount of risk, investment, and thinking 'outside of the box'. An incentive to operators is that the more business transformation that occurs the more benefits can be achieved (Venkatraman, 1994), thereby generating lucrative new revenue streams. As such, companies that are willing to take risks in adopting wireless location technology and using it creatively have tremendous possibility for achieving strategic advantage in the marketplace. However, the novelty, risk, complexity, and cost of the largest transformations and innovations will certainly prove elusive to many firms.

Summary and conclusions

Consumers and mobile professionals are demanding access to location-specific information on a wherever, whenever basis (Kivera, 2001).

This trend, fuelled by the increasing ubiquity of low-cost wireless services, growing use of GPS and other location technologies, and acceptance of the Web as a primary source of information, is compelling operators and other companies to deliver LBS to their customers. For operators, this offers a new set of revenue enhancing and differentiating value-added services.

This chapter has examined the technologies, applications, and strategic issues surrounding the development of commercial LBS. As we have seen, a variety of technologies based on the handset, mobile network, and satellite positioning systems are available, such as COO, EOTD, and AGPS. These have created the platform for a plethora of services in areas such as safety, navigation and tracking, information, and location-based transactions. Of these, safety is the key market driver, where US policy has mandated the need to determine emergency service caller location. Advertising, roadside assistance, fleet management, people tracking, road pricing and location-based products are some of the other possible LBS under development. By understanding the needs of the user, a suitable value proposition can be created that combines appropriate aspects in time and space with personal characteristics.

Mobile operators face considerable obstacles in large-scale commercialization of LBS. Key problems include standards and privacy. Before a realistic technological platform can be created, fragmented location solutions require an integrative framework. Further, privacy policies pose a significant challenge to many types of LBS. However, if such obstacles can be overcome, the strategic benefits of LBS are potentially enormous, not just in improving efficiency and effectiveness of current services, but also in developing new services and transforming core aspects of business.

In terms of research, considerable future work is needed to better understand how to leverage the value of LBS in both personal and business markets. This chapter has merely scratched the surface of an emerging and growing phenomenon. Questions for research include:

How can user value (or utility) best be modelled and created in new services?

What business models are supported by new LBS? What are the strategic benefits of LBS for firms?

How can the fragmented standards for location positioning be reconciled to provide the best service for mobile users?

Given the current relative anonymity of mobile devices, do consumers want to be identified? How can privacy be protected?

Clearly, commercial LBS for mobile consumers are still in the embryonic stages of development and use. The next few years will be fundamental in the advancement and adoption of services. Although basic services are beginning to emerge, telecommunications network advancement will enable more sophisticated location positioning services. Network standards such as GPRS, currently already rolled-out in many developed markets, are eminently suitable for LBS. Beyond GPRS, 3G network standards, offer greater flexibility; with speeds of up to 2 Mbits/s, 3G offers the ability to have simultaneous voice and data calls. With such infrastructure in place, the possibilities for new and improved services become ever larger.

Acknowledgements

An early version of this chapter first appeared as Barnes, S.J. (2003). Location-based services: the state-of-the-art. *E-Services Journal* (in press). An extended version of the chapter also appeared as Barnes, S.J. (2003). Known by the network: the emergence of location-based mobile commerce. In E. Lim and K. Siau, eds, *Advances in Mobile Commerce Technologies*. Hershey: Idea Publishing, pp. 171–189.

References

Airbiquity (2000a). The emergence of the location-commerce market. In *Proceedings of L-Commerce 2000 – The Location Services and GPS Technology Summit*, Washington DC, May.

Airbiquity (2000b). No l-commerce without l-privacy. In *Proceedings of L-Commerce 2000 – The Location Services and GPS Technology Summit*, Washington DC, May.

Barnes, S.J. (2002). Under the skin: short-range embedded wireless technologies. *International Journal of Information Management*, **22**, 165–179.

Bourrie, S.R. (2000). A sense of place: getting there from here. http://www.bluesigns/Press/bluesigns_in_news/wirelessweek_ 05152000/, accessed 15 May 2000.

Djuknic, G.M. and Richton, R.E. (2001). Geolocation and assisted GPS. *IEEE Computer*, **34**, 123–125.

Farbey, B., Targett, D., and Land, F. (1992). Evaluating investments in IT. *Journal of Information Technology*, **7**, 109–122.

Frost and Sullivan (2003). North American location-based service markets. http://www.frost.com/, accessed 25 March 2003.

Funk, J. (2000). *The Internet Market: Lessons from Japan's I-Mode System*. Kobe University, Japan: Unpublished white paper.

Greening, D. (2003). Location privacy. http://www.openmobile-alliance.org/lif/, accessed 27 May 2003.

Hamilton, T. (2000). The mobile concierge. http://www.bluesigns.com/ Press/Industry_Insights/ The_mobile_concierge.htm, accessed 1 July 2000.

Kannan, P., Chang, A., and Whinston, A. (2001). Wireless commerce: marketing issues and possibilities. In *Proceedings of 34ᵗʰ Hawaii International Conference on System Sciences*, Maui, Hawaii, January (CD-ROM).

Katz-Stone, A. (2001). Wireless revenue: ads can work. http://www.wirelessauthority.com.au/r/article/jsp/sid/445080, accessed 28 March 2001.

Kivera (2001). Kivera Spatial Suite – data sheet. http:// www.kivera.com/pdf/kls_data_sheet.pdf, accessed 27 July 2001.

Kizoom (2000). Building a WAP application: software engineering for the mobile internet. Presentation to WAP Wednesday, London, April.

Lavroff, J.L. (2000). *Location Services: How to Enhance Personal Safety and to Stimulate Lucrative Business Opportunities*. Brussels: European Commission.

Mitchell, K. and Whitmore, M. (2003). Location-based services: locating the money. In B.E. Mennecke and T.J. Strader, eds, *Mobile Commerce: Technology, Theory and Applications*. Hershey: Idea Group Publishing, pp. 51–66.

Nokia (2003). *Mobile Location Services*. Helsinki: Nokia Corporation.

O'Connor, P.J. and Godar, S.H. (2003). We know where you are: the ethics of LBS advertising. In B.E. Mennecke and T.J. Strader, eds, *Mobile Commerce: Technology, Theory and Applications*. Hershey: Idea Group Publishing, pp. 245–261.

OECD (2001). *Understanding the Digital Divide*. Paris: OECD Publications.

Puca (2001). Booty call: how marketers can cross into wireless space. http://www.puca.ie/puc_0305.html, accessed 28 May 2001.

Research in Motion (2000). The wireless workforce. http://www.rim.net/, accessed 10 August 2000.

Rockwell, M. (2003). E911: Devil is in the details. http://www.wirelessweek.com/index.asp?layout=article&articleid=CA298975&spacedesc=Departments&stt=000, accessed 27 May 2003.

Strategy Analytics (2003). *Location Based Services: Strategic Outlook for Mobile Operators and Solutions Vendors*, http://www.strategyanalytics.com, accessed 3 March 2003.

Tseng, Y.C., Wu, S.L., Liao, W.H., and Chao, C.M. (2001). Location awareness in ad hoc wireless mobile networks. *IEEE Computer*, **34**, 46–52.

Varshney, U. (2000). Recent advances in wireless networking. *IEEE Computer*, **33**, 100–103.

Varshney, U. and Vetter, R. (2002). Mobile commerce: framework, applications and networking support. *ACM/Kluwer Journal on Mobile Networks and Applications*, 7, 185–198.

Venkatraman, N. (1994). IT-enabled business transformation: from automation to business scope redefinition. *Sloan Management Review*, **35**, 73–87.

Wieland, K. (2000). Where are the location-based services? http://208.220.133.42/issues/200009/where_are_the.html, accessed 1 July 2000.

WLIA (Wireless Location Industry Association) (2003a). Wireless location technology: options for E911 and beyond. http://www.wliaonline.com/publications/globetrends.doc, accessed 27 May 2003.

WLIA (2003b). Draft WLIA Privacy Policy standards. http://www.wliaonline.com/indstandard/privacy.html, accessed 25 May 2003.

Xypoint (2001a). Platform white paper. http://www.xypoint.com/platform/tech/index.html, accessed 27 July 2001.

Xypoint (2001b). Location management. http://www.xypoint.com/platform/location.html, accessed 27 July 2001.

Zeus Wireless (2000). *Wireless Data Telemetry*. Maryland: Zeus Wireless, Inc.

Emerging wireless applications

Chapter 7

Enterprise mobility: concept and cases

Introduction

Individually, the penetration of distributed networks and mobile telecommunications in the developed world have evoked considerable change in our daily lives – how we work, live, and learn. Combined, the potential impact of the wireless networks and related applications is immense. Enterprise mobility appears to be one of the prime candidates for early adoption benefits, and forms the basis of this chapter.

Although the literature on commercial wireless applications has predominantly focused on B2C markets, following the patterns in the media and e-commerce research, it is now becoming clear that mobile networking will provide a tremendous impetus to the development of other strategic applications for businesses. It is now becoming clear that the impact of mobile computing and m-commerce goes much further; wireless technologies have the potential to transform activities both within and between businesses (Alanen and Autio, 2003). Industry sources predict that corporate demand is likely to drive the wireless market forward and many applications are being developed for wireless enterprise computing (Wrolstad, 2002). Indeed, by 2004, cost savings could permit wireless business services around the world to generate annual value of up to $80 billion, and at least as much value could be created if corporations used wireless services to improve their current offerings or to deliver new ones (Autio *et al.*, 2001).

This chapter picks up on the emerging area of wireless applications in the business. It provides a background to some of the conceptual ideas of enterprise mobility, and begins to apply these in a number of original case studies. The next section examines some of the developments in enterprise mobility, detailing a conceptual framework for understanding the potential of mobile data communications in the B2E space. The chapter continues by outlining a brief survey of a number of wireless applications currently being used in New Zealand business. This is followed by some recent case studies, analysed using the conceptual framework. The chapter rounds off with a summary and some conclusions regarding the future of enterprise mobility.

An introduction to enterprise mobility

Wireless data communications can provide significant business benefits for corporate infrastructure and a large number of corporate solutions have been developed to this end (Barnes, 2002a). Messaging is one key area of application. For example, advances in wireless messaging allow mobile workers to direct specific incoming messages to specific devices; such control helps mobile workers direct urgent e-mails to handheld devices or cell phones, and less urgent matters to secondary devices such as desktop PCs (Ferguson, 2000). Other mobile office tools are also available – linking to fax, databases, schedules, and file transfer (Arthur D. Little, 2000; Research in Motion, 2000).

On a much broader level, wireless networks and devices can help to strongly integrate remote, disparate, or roaming employees into the corporate infrastructure. These include functionally disparate or mobile employees, such as salespersons, and remote workers, such as integrating geographically dispersed units in a travel agency in Australia (Arthur D. Little, 2000). Thinking more generally about the mobile workforce, employees are enabled to work in their virtual office at any time, in any place, and anywhere. Mobile devices and data connections can provide important links to company networks and systems that are key to the effective performance of work (IBM Pervasive Computing, 2000a, b). This is demonstrated in Figure 7.1. In the example, the Blackberry is an additional channel for data communication, but it provides access to a variety of corporate systems such as customer relationship management (CRM), sales force automation (SFA), groupware, the Web, and even some supply chain

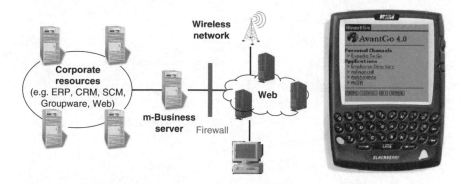

Figure 7.1   Enterprise mobility – accessing the corporate resources

management (SCM) and enterprise resource planning (ERP) functions. Ovum predicts that the number of mobile-office users will grow
to 137 million by 2004, up from 15 million in 2000 (Autio *et al.*, 2001).

A framework for understanding enterprise mobility

Figure 7.2 shows a conceptual framework for understanding the
potential development of enterprise mobility in organizations,
which we will refer to as the Mobile Enterprise Model (MEM). The
diagram shows three axes, each with three possible positions. The
axes are mobility, process, and market. Briefly, the axes can be
described as follows:

Mobility. This describes the level of geographic independence of enterprise workers, enabled by the wireless data solution. The first level is
'transient', describing the basic support of employees as they move
from one location to another. These employees are geographically tied
to the locations between which they move. The second level is 'mobile'.
Here, employees have a much higher degree of geographic independence from the enterprise for prolonged periods of time, but they
inevitably return to corporate locations to perform certain functions.
Finally, the highest level of mobility is 'remote'. At this level, employees
are almost completely removed from the corporate location, being
empowered with a very high degree of geographic independence.

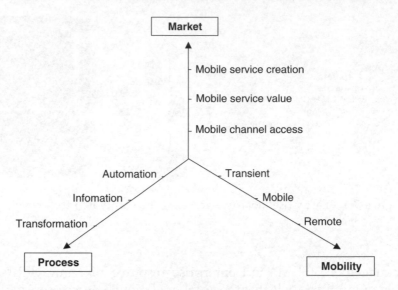

Figure 7.2 Dimensions and stages of mobile enterprise – the Mobile Enterprise Model (MEM)

Process. This describes the change in work configuration and processes. The first level, 'automation', refers to efficiency gains in existing processes transferred to the mobile data environment. 'Infomation' brings in a degree of effectiveness and knowledge work gains via the mobile solution. Finally, 'transformation' describes a fundamental degree of change in organizational processes using the mobile medium. At this level, the nature of work and job roles may be transformed by the mobile medium.

Market. This describes the value proposition in the marketplace; typically, it refers to the changes in products, services, and relationships with customers, but it may also contain market experiences with suppliers and business partners. Again there are three levels. At the lowest level, 'mobile channel access', the mobile medium is being used largely as a conduit for information for mobile employees, without significantly different services. At the intermediate level, 'mobile service value', the wireless solution is being used to add significant value to the market offering. Thus, there are specific areas where the product or service level is being significantly enhanced using enterprise mobility. Finally, at the highest level, 'mobile service creation', the wireless medium is being used to create entirely new service offerings or products.

Accordingly, we can envisage three major phases in the use of enterprise mobility in organizations:

Phase I: mobile employee linkage. This phase of enterprise mobility emphasizes entry-level enterprise offerings. It focuses on establishing the appropriate wireless infrastructure to 'link-in' transient employees to allow access to corporate data and aid in efficiency for existing work.
Phase II: mobile employee empowerment. In this phase, the work patterns of employees are driven by the availability of corporate knowledge on the mobile medium. In this way, mobile employees are able to significantly improve the effectiveness of work configurations and therefore of the products or service provided.
Phase III: mobile enterprise creation. Only in this highest phase of enterprise mobility can the organization boast truly mobile employees and services. At this level, employees can exist separately of the geographic constraints of an organization, supported by wireless solutions. The nature of work has been significantly transformed to take advantage of the new environment, and the roles of individuals are likely to be very different. In addition, the mobile enterprise is able to offer new and different products and services than before supported by the mobile paradigm.

Figure 7.2 is explored further later, where it is used as a framework for analysing the development of enterprise mobility in a number of recent case studies.

A survey of practice: New Zealand businesses with developed mobile offerings

One market where a considerable number of enterprise wireless applications are being implemented is New Zealand. Although only a relatively small market, New Zealand has the benefit of being a developed economy with a relatively advanced technological base. Typically, many companies in the technology arena have used New Zealand as a test-bed for new applications or technologies. Such companies include Microsoft, HP, Ericsson, and Lucent. In this section, we will provide a brief survey of the current areas of application of wireless in businesses in New Zealand, shown in Table 7.1.

The pattern of adoption of wireless technologies in businesses in New Zealand is diverse, ranging from banking to car auctions.

Table 7.1 A sample of wireless applications in use in New Zealand business

Company	Business area	Application area	Focus	Primary					Support			
				IL	Op	OL	MS	Sv	AI	HR	PD	Pr
ASB FastNet	Bank	Wireless banking	S	o	o	o	o		o			
Courier Post NZ	Logistics	Dispatch and tracking system	S, Ps		o	o			o	o		
Delta	Utilities management	Supports data management	S, Ps							o	o	
Green Acres	Home services	Dispatch, customer management	S, R, Ps	o	o					o		
Kapiti Cheeses	Manufacturing	Inventory management, SFA	Pd, R	o	o	o	o		o	o		
Mainfreight NZ	Logistics	Dispatch and tracking system	S		o	o				o		
MasterPet Foods	Distributor of pet food	Inventory management, SFA	Pd, S, Ps	o	o	o	o					
Nat'l Road Carriers	Logistics	Vehicle tracking and monitoring	S, Ps	o	o		o		o		o	
National Bank of NZ	Bank	Wireless banking	S	o	o		o		o			
Northland Health	Health authority	Message broadcasting	S		o		o		o			
NZ Funds Mgt.	Financial services	Customer management, reporting	S, R		o		o		o			
NZ Stock Exchange	Stock exchange	Message broadcasting	S		o		o					
One Source	Equipment service	Service support, equipment mgt.	S, R, Ps	o	o		o			o		o
Quotable Value	Property valuation	Facilities reporting and input	S, Ps	o	o		o		o			
Relocations	Relocation	Inventory cataloguing, SFA	S, R		o			o	o	o		
Service Printers	Printers	Support, SFA	S, R					o		o		
Shell NZ	Petrol sales	Business information service	Pd, R		o		o		o			
TMS Group	Financial services	Message broadcasting	S, R		o		o					
Transfield	Engineering	Facilities management	S, R		o		o					
Turners Car Auctions	Car auctions	Information service, SFA	S, Ps	o			o		o			o

Key: Focus: S = service; Pd = product; R = relationship; Ps = process; Value chain activities: IL = inbound logistics; Op = operations; OL = outbound logistics; MS = marketing and sales; Sv = servicing; AI = administrative infrastructure; HR = human resources; PD = product development; Pr = procurement
Source: Barnes (2002b); Sewell (2002); Telecom NZ (2002k)

Table 7.1 provides a brief overview of a number of companies that are at the forefront of mobile business. Table 7.1 uses Porter's value chain (Porter and Millar, 1985) as a tool for analysing the impact of the specific applications on the business, as advocated by Barnes (2002b). It also provides some indication of the focus of the application on products, services, relationships, or processes.

It is interesting to note several things about the pattern of applications being used in the businesses shown in Table 7.1. Message broadcasting is a popular application being used by a number of organizations, building on the success of text messaging in the B2C space. Vehicle tracking is another popular application, focusing particularly in the logistics industry. Aside from these, the overwhelming emphasis is on field-force automation, such as in service and sales. It is interesting to note the number of banking and financial services applications shown in Table 7.1.

In terms of the impact of the wireless applications, it is clear that the main focus is on improving the product or service offering, and building better relationships, rather than major transformation of processes. The primary activities in Porter's value chain most impacted by wireless systems are operations, and marketing and sales, the latter being particularly for mobile SFA/CRM systems. Logistics is also an area where there is transformation for some companies. For support activities, the infrastructure component is the biggest area of impact, as the new wireless infrastructure is built. Human resources follow this as a key area for wireless systems, especially since many of the applications focus on field-force automation.

The next section focuses on five of these case studies in more detail, analysing the impact of the wireless solutions on enterprise mobility.

Case studies of wireless applications in New Zealand business

In this section, we examine in more detail five of the wireless applications from the survey above. These cases were selected for several reasons. First, they are all examples of enterprise mobility in the B2E space. This is a key focus of this research and of the Mobile Enterprise Model. Second, they are case studies that are being used by one of the major telecommunications companies, Telecom New Zealand, as exemplars of B2E mobility. Finally, initial contact had already been made with several of the case organizations and systems developers regarding data collection.

The results of these case studies are based on a mixture of primary and secondary research. In particular, they are based on semi-structured interview, short video interviews from Telecom New Zealand, internal reports, public relations documents, and personal and panel discussions at the Wireless Expo organised by Telecom New Zealand in September 2002.

A summary of the case studies is given in Table 7.2. This provides a brief description of the case organizations, along with some summary data, the focus of the B2E application, and where benefits are being obtained.

DELTA Utility Services

DELTA's business revolves around the management of physically distributed assets in the electricity industry. The mobile enterprise system developed by DELTA allows remote access to its geographical information system (GIS). Developed in 2000, the GIS is a comprehensive and information-rich mapping system that enables DELTA to record and track the presence and technical specifications of all the assets it manages across New Zealand. The system is based around a laptop equipped with GIS software, complete with a mobile data card. Field personnel, typically (electricity) network designers, are able to access the DELTA server via a secure virtual private network (VPN). The mobile enterprise system allows fast transmission of rich map data to field personnel. Data can be queried, mapped, and updated in the field, which is then audited in the main office and posted to the database. A trace function allows staff to check a circuit for network connectivity.

The value proposition of the mobile enterprise system is largely based around the improved accuracy and timeliness of information. The paper-based maps in the previous system were somewhat limited, and only useful as a guide, containing little detail on the specific assets. In operational terms, the wireless system also provides reductions in travel times for remote staff. The efficiency and accuracy of the new system has knock-on effects for customer service;

Table 7.2 Summary of mobile enterprise case studies

Company	Annual revenues	Staff	Description of application	Main benefits focus	Specific benefits
Delta Utility Services Asset management and contracting company for Dunedin Electricity Limited, New Zealand's fifth largest electricity distribution company	Manages assets of NZ$230m+	350	Remote field force support. Real time GIS data on laptops	Efficiency and accuracy	Rich GIS data checked on site Work integration of remote staff Accurate, real-time data access Reduced data entry Enhanced paperless security Improved customer service
Turners Auctions Vehicle inspectors and New Zealand's largest car auctioneers, with 90% market share	NZ$400m	200	Field force automation. Mobile vehicle inspection data collection using an iPAQ and digital camera	Efficiency and service gains	Accurate, real-time data collection Increased productivity of staff Reduced data entry Improved customer service Reduced overheads Competitive advantage
Onesource Office products and servicing of office equipment	NZ$100m+	500+	Field force automation. Remote job allocation and management using iPAQ. Barcode scanning of parts	Productivity and service gains	Increased field force productivity Improved customer service Reduced costs Increased data accuracy Better management control
Green Acres Home services provider, including cleaning, gardening, plumbing and electrical, and pet care	NZ$30m	600 sub-franchises	Field force automation. Remote job allocation and management using iPAQ. Customer relationship management	Superior customer service	Increased field force productivity Improved customer service Shorter response times Reduced costs, phone charges Relationship building/sales leads
Masterpet Foods Limited New Zealand's leading supplier of pet accessories and pet food to pet stores, veterinarians, and supermarkets, with a 75% market share	NZ$40m+	150	Sales force automation. Inventory, pricing, customer history, and order placement using an iPAQ	Efficiency and accuracy	Instant customer information/reports Reduced order processing time Reduced data entry Efficiency in inventory management Improved sales productivity Work integration of remote staff

Source: Telecom NZ (2002a–e).

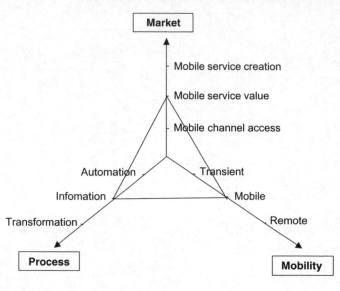

Figure 7.3 MEM analysis of DELTA

accurate, up-to-date information is directly related to optimal maintenance and asset replacement recommendations for customers (Telecom NZ, 2002a, f).

In terms of the Mobile Enterprise Model, DELTA has moved into the mobile employee empowerment phase (see Figure 7.3). The mobile data solution for electricity network designers has enabled them to access critical information and perform activities that go beyond efficiency to effectiveness and decision-making. The designers can work in the field for prolonged periods of time without returning to the main office. In terms of the market offering, the mobile solution has enabled enhanced recommendations for optimal maintenance and asset replacement.

Turners Auctions

Turners Car Auctions have applied enterprise mobility in the area of vehicle inspections. Turners typically inspect vehicles for clients in the lease sector, to enable them to make decisions regarding repair or sale of vehicles. The inspectors began using the wireless technology in May 2002. The field equipment consists of a handheld device with a wireless data card and software from Net Result. This is used to access the Turners Vehicle Inspection intranet on the corporate LAN via a secure IP network. Photographs of a vehicle being inspected are taken with a digital camera, complete with a flash card. This data can then be transferred via the flash card to the iPAQ.

The wireless system eliminates the inefficiencies associated with double data handling. The previous system relied on typing-up paper-based reports, using forms specific to each client. Digital pictures were previously uploaded to PC via USB cables. Information captured in real-time are compiled into online reports delivered via a customer extranet, usually to a lease manager. Typically, this enables the client to select the auction or damaged vehicles sections of Turners, for further business. The new process also reduces freight costs for customers – since vehicles are not required at a central inspection point. Customer service has also improved, and the work for one major lease client has increased by 150 per cent as a result (Telecom NZ, 2002b, g).

Figure 7.4 shows the degree of enterprise mobility. The enterprise mobility of Turners Auctions is very much focused on efficiency gains and automation, rather than effectiveness and knowledge gains. As such the work configuration aspect is less developed.

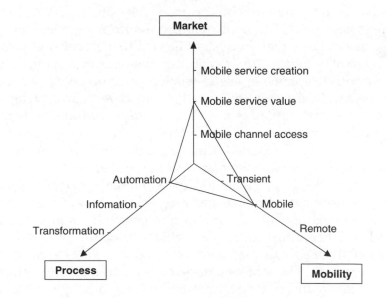

Figure 7.4 MEM analysis of Turners Auctions

However, the field inspectors can exist independently of the office for prolonged periods of time, especially since they are not required to return to complete inspection reports. The availability of real-time information, which can be immediately used to get quotes for repair, for example, is a major improvement to the service offering.

Onesource

Onesource have used enterprise mobility to connect to its mobile service technicians, and the system is currently being piloted in Auckland. In all, the organization's 180 technicians service over 26,000 pieces of office equipment in New Zealand. The system is based on the use of a handheld computer (an iPAQ with a rugged case and second battery sleeve) with a wireless data card and barcode scanner. The serial number of a piece of equipment, such as a photocopier, is a key piece of data in the job management system. When a customer calls in, the job is automatically linked to the area technician, who is intimately familiar with each machine's service history and the customer's requirements. Once the job is accepted, they are sent the details of the job. If the preferred technician does not respond, a separate message is sent to their cell phone after 15 minutes, before the job is passed on to the 'next best' person (Telecom NZ, 2002c, h).

The system has led to noticeable process improvements for staff. Onesource indicate that the system has led to a 30 per cent increase in productivity and a 20 per cent improvement in response times. The previous telephone-based despatch process involved on average three messages sent via pager or cell phone to the nearest technician. Inaccuracy was common, as the technician recorded the 15-character part numbers on paper and then relayed this over the phone to the dispatcher, who then entered it into a computer. The PDA software includes data entry parameter checks to reduce errors.

Alongside the implementation of a mobile enterprise system, work practices have become more competitive. Management are now able to extrapolate performance information from the technician's PDA, including travel times (which are significantly reduced), the reliability of specific machines, and response times to customers' service requests. The system empowers staff to plan and prioritize their workload and engenders competitiveness among service staff, aiming to provide improvements in service.

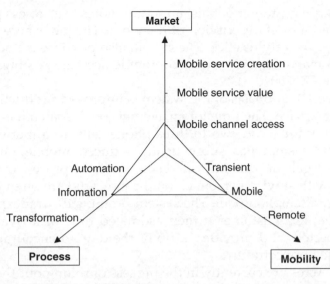

Figure 7.5 MEM analysis of Onesource

It is estimated that investment in the system will pay back in 14 months. Onesource are currently in the process of the next phase of implementation, deploying the system to 90 handhelds. Typically, five technicians are being trained in a day, every few days (Telecom NZ, 2002c, h).

Figure 7.5 shows the degree of enterprise mobility. As demonstrated in this section, the workforce is empowered both to work geographically independent from the office for long periods and to use information to change their performance. However, aside from the efficiency and response improvements, the market impact is minimal.

Green Acres

Enterprise mobility has been applied in the Trade Services Group of Green Acres. This group, launched in 2002, provides electrical and plumbing services. Electricians in the Auckland area are currently piloting the mobile job management system (JMS). Field service staff access the JMS using a handheld computer (an iPAQ with a rugged case and second battery sleeve) with a wireless data card. The system guides staff through a prompted, step-by-step process that provides full customer details, important notes (e.g. access points to a site), and a brief description of the work. The job is accepted and completed by a stylus click. The system also provides a stock inventory and management system to provide accurate costing of parts (Telecom NZ, 2002d, i).

The overall emphasis of the system is improved productivity and customer service. The time for an average job despatch has been cut from 10 minutes, and two to three phone calls, to a matter of seconds. The system has subsequently reduced mobile phone call charges, largely at peak rates, and simplified the process of communicating with service personnel 'on the job', who are often too busy to answer a telephone call. The system prompts the tradesperson to phone the next customer if their estimated time of arrival is later than expected, and provides a 'tip of the day' to encourage cross- and up-sell opportunities.

Green Acres are currently in the process of rolling out the system nationwide in the Trade Services group. They are also in the process of implementing a facility to print invoices on the job and a swipe-card instant payment option.

Figure 7.6 provides an examination of the enterprise mobility of Green Acres. Due to the nature of franchised operations, the system

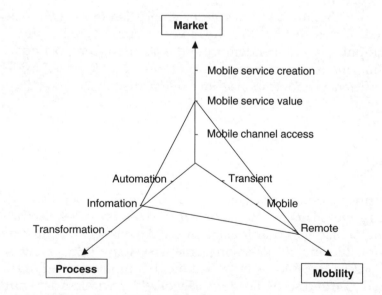

Figure 7.6 MEM analysis of Green Acres

assists tradespersons in operating almost entirely remotely. The nature of information goes beyond simple efficiency to actually improve the effectiveness of operations and add to service value in areas such as customer knowledge and sales opportunities.

Masterpet

Masterpet was already relatively advanced in the area of sales force automation before the introduction of the new mobile sales force solution. The sales force were already using large touch screen TX tablets and dialup modems to send and receive product and ordering data. However, these tablets were bulky, expensive (NZ$10,000 each), and were prone to damage of moving parts (especially harddrives) as sales representatives carried them from customer to customer. The new system is based on an iPAQ linked by Bluetooth to a mobile phone data connection, fitted with a wireless data card and field sales software. The aVya software allows sales representatives to perform a variety of common tasks using a handheld computer. This includes placing an order, quotes, generating price lists, updates on the latest special offers, current stock availability, and standard customer information (including, e.g., credit limits). Invoices are created automatically, and salespersons can also generate sales reports and demand information without the need for further data entry at the office. Access to e-mail and personal information management (PIM) functions, such as to-do lists, are also provided.

Orders are instantly fulfilled. Orders entered into a handheld by a sales person at a store can be picked and packed and ready to deliver in 10 minutes. The process works as follows (Telecom NZ, 2002e, j):

> Orders are taken via an iPAQ and aVya application.
> The order is sent over the mobile IP network.
> Orders are uploaded to the Navision ERP system.
> Orders are released to the voice activated picking system (VOCollect).
> Orders are picked and packed. Confirmation is sent to the ERP system.
> The ERP system posts and generates the invoices, which are sent electronically via the Internet.
> Packing slips and invoices are printed.
> Goods are shipped to the customer.

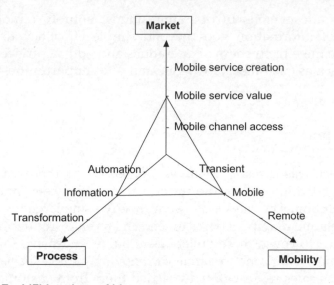

Figure 7.7 MEM analysis of Masterpet

More than half of the sales force is currently using the mobile system, and this is currently being rolled out to all 30 sales representatives in November 2002. Accuracy has increased very considerably using the system, with the error rate falling from around 12 to just 2 per cent. This equates to a NZ$60 per order cost saving for errors. Typically, this means that sales support staff focus more on adding value to the service rather than correcting inaccurate orders. Improvements in productivity are also apparent, and salespersons typically make at least two extra sales visits per week, or have extra time to build customer relationships.

Figure 7.7 depicts the degree of enterprise mobility at Masterpet on the MEM. The mobile solution affords a high degree of mobility for sales representatives, and numerous functions, such as invoicing and reports, can be completed from the handheld device. Although there are significant efficiency benefits, the impact of the solution goes well beyond this to impact the effectiveness of sales personnel and shifting the nature of work more towards customer relationship-building. As a result, the customer experiences an enhanced level of service value supported by the mobile information system.

Conclusions

Wireless technology has provided the platform for a vast array of mobile applications in the enterprise space. This chapter has examined the impact of wireless solutions on B2E enterprise relationships. After a brief background on enterprise mobility and a snapshot of some of the companies using mobile enterprise solutions in New Zealand, the focus has been turned towards a number of very recent case studies of enterprise mobility. The five cases studied – DELTA, Turners, Onesource, Green Acres, and Masterpet – have all successfully implemented mobile solutions in the business-to-employee space in early 2002. The benefits of the solutions are significant, sometimes strikingly so. Nevertheless, the impact of the solutions, although sharing some common characteristics, varies very considerably. To analyse the development of enterprise mobility and to benchmark the status of the concept in each organization, the research has employed the Mobile Enterprise Model. This provides a simple assessment of any organization in three aspects of enterprise mobility: mobility, process transformation, and market value. The application of MEM to the case studies provides some clear demonstrations of the differences in B2E m-enterprise solutions. Overall, the emphasis of the cases is on Phase II mobile enterprise developments – mobile employee empowerment. Each of the cases goes beyond basic mobile employee linkage, but none of them move toward the advanced levels of mobile enterprise creation.

This area of research is clearly very embryonic. The cases studied are state-of-the-art applications of wireless technologies for enterprise mobility. As such, the framework provided gives some early directions for examining the conceptual idea of enterprise mobility, but not a fully formed theoretical offering. Future research will help to provide a more detailed and grounded theoretical framework for analysing enterprise mobility. This is likely to build on further developments in this fast-moving field of application and research.

Time will tell which wireless technologies or applications become pervasive and dominant. However, it is clear that enterprise mobility has an important part to play in the IT strategies of many organizations in the future; the applications presented in this chapter are being successfully used or piloted in organizations today. With further technological advances, we are likely to encounter some even more creative and interesting applications in the enterprise.

Acknowledgements

This chapter first appeared as Barnes, S.J. (2003). Enterprise mobility: concept and examples. *International Journal of Mobile Communications*, in press.

References

Alanen, J. and Autio, E. (2003). Mobile business services: a strategic perspective. In B.E. Mennecke and T.J. Strader, eds, *Mobile Commerce: Technology, Theory and Applications*. Hershey: Idea Group Publishing, pp. 162–184.

Arthur D. Little (2000). Serving the mobile customer. http://www.arthurdlittle.com/ebusiness/ebusiness.html, accessed 15 January 2000.

Autio, E., Hacke, M., and Jutila, V. (2001). Profit in wireless B2B. *McKinsey Quarterly*, No. 1, 20–22.

Barnes, S.J. (2002a). Unwired business. *E-Business Strategy Management*, **4**, 27–37.

Barnes, S.J. (2002b). The mobile commerce value chain: analysis and future developments. *International Journal of Information Management*, **22**, 91–108.

Ferguson, K. (2000). Messaging service focus on mobile workers. http://www.forbes.com/, accessed 6 September 2000.

IBM Pervasive Computing (2000a). CRM/ERP/SCM. http://www-3.ibm.com/pvc/mobile_internet/crm_erp_scm.shtml, accessed 28 November 2000.

IBM Pervasive Computing (2000b). Extending SAP systems to pervasive computing devices. http://www-3.ibm.com/pvc/tech/sap.shtml, accessed 27 November 2000.

Porter, M. and Millar, V. (1985). How information gives you competitive advantage. *Harvard Business Review*, **63**, 149–160.

Research in Motion (2000). Building a business case for wireless. http://www.rim.net/, accessed 15 November 2000.

Sewell, J. (2002). *M-business*. Unpublished report. Victoria University of Wellington.

Telecom NZ (2002a). *Telecom More Mobile and DELTA*. Telecom Mobile Worker, Series No. 66664, Wellington: Telecom NZ.

Telecom NZ (2002b). *Telecom More Mobile and Turners*. Telecom Mobile Worker, Series No. 66680, Wellington: Telecom NZ.

Telecom NZ (2002c). *Telecom More Mobile and Onesource*. Telecom Mobile Worker, Series No. 66672, Wellington: Telecom NZ.

Telecom NZ (2002d). *Telecom More Mobile and Green Acres*. Telecom Mobile Worker, Series No. 66688, Wellington: Telecom NZ.

Telecom NZ (2002e). *Telecom More Mobile and Masterpet*. Telecom Mobile Worker, Series No. 66696, Wellington: Telecom NZ.

Telecom NZ (2002f). *Telecom More Mobile and DELTA*. Promotional Video. Wellington: Telecom NZ/DELTA.

Telecom NZ (2002g). *Telecom More Mobile and Turners*. Promotional Video. Wellington: Telecom NZ/Turners.

Telecom NZ (2002h). *Telecom More Mobile and Onesource*. Promotional Video, Wellington: Telecom NZ/Onesource.

Telecom NZ (2002i). *Telecom More Mobile and Green Acres*. Promotional Video. Wellington: Telecom NZ/Green Acres.

Telecom NZ (2002j). *Telecom More Mobile and Masterpet*. Promotional Video, Wellington: Telecom NZ/Masterpet.

Telecom NZ (2002k). *Mobile Times*, August–September, Wellington: Telecom NZ.

Wrolstad, J. (2002). New wireless tech pushes mobile sales apps. http://wirelessnewsfactor.com/perl/story/19888.html, accessed 6 November 2002.

Chapter 8

Wireless digital advertising

Introduction

In the last 10 years, the Internet has witnessed extraordinary market penetration, and it continues to grow unabated; from approximately 500 million users in 2001, the Internet is predicted to grow to approximately 1 billion users by 2004 (IDC Research, 2001). One area that has experienced tremendous growth is online advertising, a global market predicted to grow from approximately $10 billion in 2001 to over $28 billion by 2005 (Gluck, 2001). In essence, online advertising includes any paid advertisement, from banners to sponsorships, which appears on the Web or other Internet channels, including e-mail. Online advertising has become increasingly sophisticated as retailers and service providers utilize the Internet's capabilities to focus on prospective consumers, identify their particular interests, and position products and services in a way that not only leads to one-time purchases, but to long-term, ongoing customer relationships (Dutta and Segev, 2001). This potential has been improved by the development of a range of online customer relationship management (CRM) tools and techniques, whereby online advertisements can be electronically linked to back-end systems that facilitate interactive e-mail campaigns, incremental customer profiling, opt-in mailing lists, special offers, and many other opportunities (Adflight, 2000).

Online advertising provides significant benefits over traditional media, providing the first advertising medium that is truly measurable, accountable, and targeted for one-to-one marketing (Meeker, 1997). Indeed, marketing professionals can actually measure how many prospects have seen or responded to an advertisement, and achieve more precise targeting of advertising based on, for example, geography, demographics, or similar profiling, by placing specific ads on particular sites (Eighmey, 1997). Such advertisements can be typically more interactive and personal (Lot21, 2001). Online media also has inherently lower costs than traditional media, such as TV, with the possibility of using tools for continuous streamlining, improvement, and optimization. Recent evidence also suggests that online advertising is having a major impact on offline buying patterns; for example, many buyers who purchase a car offline have first researched the product online (Adflight, 2000). With such benefits, it is perhaps not surprising that, despite the economic slowdown, online advertising is still relatively buoyant, accounting for 3 per cent of all advertising in 2001, and predicted to rise to 8 per cent by 2005 (Gluck, 2001).

The convergence of the Internet and wireless telephony has also presented a new platform for advertising, and, if analysts are correct, one that is potentially much stronger than the wired Internet. Using devices such as mobile phones and PDAs, and based on platforms such as the WAP, SMS, and variants of HTML, global wireless advertising revenue is predicted to grow from $750 million in 2001 to $16.4 billion by 2005 (Ovum, 2000). Europe, the Asia-Pacific, and North America are expected to be the core markets, with wireless advertising revenues of $5.98, $4.71, and $4.56 billion, respectively, in 2005 (Ovum, 2000).

Wireless devices have a number of features that together form a fertile bed for advanced wireless services, including advertising (Kannan *et al.*, 2001). Typically, devices are very personal to the user, and carried on the person; aspects of the context of the user, such as time and place, can be measured and interpreted; services can be provided at the point of need; and applications can be highly interactive, portable, and engaging. For the consumer, wireless advertising can be very personal, timely, and relevant, or even integrated with other services, such as games, in a near-seamless way; for the advertiser, wireless ads can be accurately targeted and measured (Katz-Stone, 2001).

The purpose of this chapter is to examine the nature and implications of this emerging channel of advertising. The next section explores the nature of advertising services in this new medium, including some examples and recent experiences. The third section examines a simple research model for mobile advertising. Finally, the chapter rounds off with some conclusions, and provides some predictions on the future of wireless advertising.

Advertising on the wireless Internet

The wireless Internet presents an entirely new advertising medium that must address traditional marketing challenges in an unprecedented way (Windwire, 2000). Whereas traditional marketing channels have guidelines for successfully communicating with customers, gaining attention and recognition, this process cannot translate unchanged into wireless marketing. New approaches must be creative and engaging to take full advantage of the wireless medium (Kong, 2001).

Key industry players in the value chain providing wireless advertising to the consumer are agencies, advertisers, wireless service providers (WSPs), and wireless publishers. For agencies and advertisers, the wireless medium offers advanced targeting and tailoring of messages for more effective one-to-one marketing. For the WSP, the gateway to the wireless Internet (e.g. NTT DoCoMo, AT&T and TIM), wireless advertising presents new revenue streams and the possibility of subsidizing access. Similarly, wireless publishers (e.g. the *Financial Times, New York Times*, and *CBS Sportsline*), as a natural extension of their wired presence, have the opportunity for additional revenue and subsidizing access to content. At the end of the value chain, there is potential for consumers to experience convenient access and content value, subsidized by advertising (Windwire, 2000). Furthermore, value-added marketing messages can be targeted to preferences, lifestyle, and location (Lot21, 2001). However, what is unclear, as yet, is how consumers will respond to the idea of wireless advertising. Although, as we shall see below, pilot studies have shown some positive response, particularly from younger age groups, the results are quite limited and certainly not generalizable.

The industry players are still in the very early phases of wireless advertising, largely based on experimentation. However, some general lessons are being learnt. The purpose of this section is to examine some of the core types and recent experiences of wireless advertising. Like the wired medium, advertising on the wireless Internet can be categorized into two basic types: push and pull. These are illustrated in Figure 8.1. Push advertising involves sending or

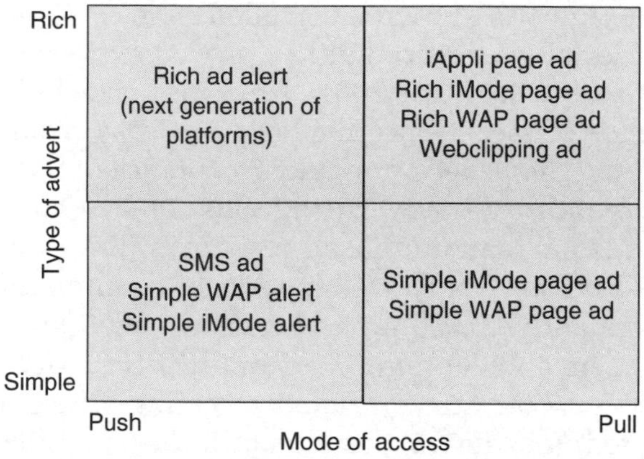

Figure 8.1
Categorization of wireless advertising – with examples

'pushing' advertising messages to consumers, usually via an alert or SMS text message. Pull advertising involves placing advertisements on browsed wireless content, usually promoting free content. Typically, the available platform for advertising dictates the complexity of the advertisement, with SMS being simple and text-based, and iAppli allowing complex use of graphics and Java. Let us continue by examining this typology in more detail.

Wireless push advertising

Push advertising is currently the biggest market for wireless advertising, driven by the phenomenal usage of SMS. Some 4–500 billion messages were sent globally in 2001 (Frost & Sullivan, 2001; Kong, 2001). The Wireless Advertising Association standard for SMS ads ranges from 34 to 160 characters (WAA, 2001). The range of products and services currently advertised is immense. Examples include a bulletin service for a chain of nightclubs in Ireland, a dating service, and a Europe-wide Euro2000 football messaging service (Puca, 2001).

The idea of an advert being sent directly to an individual's phone is not without legislative concerns. Typically, common sense dictates that push marketing should be reserved for companies who have an established relationship and permission to push wireless communications to users. Indeed, privacy and consumer rights issues do restrict the use of push advertising, leading to the promotion of 'opt-in' schemes. In essence, 'opt-in' involves the user agreeing to receive advertising before anything is sent, with the opportunity to change preferences or stop messages at any time.

An analysis of SMS usage has shown unrivalled access to the age group from 15 to 24 years – a group that has proved extremely difficult to reach with other media (Puca, 2001). Key reasons include privacy, flexibility, absence of face-to-face contact, and easy availability of the SMS medium. The penetration and acceptance of SMS provides an interesting platform for advertising. Studies by Quios and Ericsson found surprisingly favourable ad responses (Ericsson, 2000; Quios, 2000). The Ericsson study was based on a sample of approximately 5000 users and 100,000 ad impressions; the Quios study examined 35,000 users and 2.5 million ad impressions. More than 60 per cent of users on both surveys liked receiving wireless advertising. Reasons cited for the favourable attitudes to SMS advertising include:

Content value. People participating in these trials were open to advertising as long as there is a free service or other inducement. In the Ericsson study users were given free text-messaging capabilities.

Immersive content. SMS messaging has an interactive quality that is particularly attractive to younger users. This includes, for example, are messages in response to an SMS request.

Ad pertinence. At the present, there is little in the way of junk messaging or 'spam'. Since advertising is less common, it tends to be highly pertinent, such as new rates for services or other offers of immediate relevance. Related to this, ad recognition was very high. For example, in the Quios study, 79 per cent of participants had at least 60 per cent recall of wireless advertising.

'Wow' factor. Since there is an absence of clutter, messages maintain a degree of surprise. Users are not yet inured to such ads and so tend to have a high degree of acceptance.

Viral marketing. Related to the high impact of advertisements and usage of mobile phones for social communication, community is an important aspect of viral marketing via wireless devices. For example, 70 per cent of users in the Quios study recommended the Euro 2000 messaging service to a friend.

Personal context. Some services can be personalized, based on the user's preferences. This was a particularly favourable aspect of the Ericsson study. In the future, location-aware technology can extend this to even allow advertising at the point-of-sale.

Two other recent studies of SMS advertising in the United Kingdom found similarly positive results (Enpocket, 2002, 2003).

The first study of 5000 consumers by Enpocket (2002), all of whom had given permission to receive third-party marketing, found a very high level of acceptability for SMS advertising from trusted sources (63 per cent). The results demonstrated that consumers were resistant to SMS marketing until exposed to it, but then found that it was as acceptable as TV or radio advertising (68 and 65 per cent, respectively), and more acceptable than most other media (e.g. 43 per cent for magazine inserts, 27 per cent for direct mail, and 9 per cent for telesales). However, although consumers were comfortable with ads when the media and permission holder is the carrier or portal, they were much less accepting of advertising from other sources (with only 35 per cent of respondents finding this acceptable).

The second study by Enpocket (2003), which examined over 200 SMS campaigns between October 2001 and January 2003 and interviewed more than 5200 consumers, found a high level of virality and consumer response for SMS ads. The data suggest that on 94 per cent of occasions recipients read SMS ads, and some 23 per cent of those surveyed showed or forwarded an ad to a friend. Furthermore, on average, 15 per cent of individuals responded to SMS ads, with a response rate of 46 per cent for the most successful ad campaigns. Typically, responses were in the form of replying to a message (average 8 per cent; best 27 per cent), visiting a Web site (average 6 per cent; best 19 per cent), visiting a store (average 4 per cent; best 15 per cent), or buying the advertised product (average 4 per cent; best 17 per cent). Enpocket (2003) concluded that responses for SMS marketing were between 2 and 15 times as effective as direct mail (5.2 per cent response) and e-mail (6.1 per cent response), pointing out that the greatest value of the medium is in terms of the personal environment of the mobile device, the high standout and novelty of ads, ease of response, and the ability to target ads based on a combination of demographics, time, location and lifestyle interest/ context to increase the quality and relevance of what is offered.

Further expansion of push-based advertising could change some of these perceptions. Although the novelty value was positive for these trials, the corollary is that growth in advertising can dilute the advertising message or even annoy users (Hamblen, 2000).

Other services, such as WAP and iMode, can offer advertising alerts, although this is still in the early stages (WestCyber News, 2001). Currently, trials of WAP e-mail ads are being conducted around the globe. A WAP alert involves the user being sent a short e-mail with a link to a WAP page. However, until standards like GPRS are widespread, WAP users are required to dial-up for a connection to receive these. This makes it critical that advertising is highly targeted and relevant to the consumer; the user should not believe that viewing the advertiser's message is wasting expensive airtime (Lot21, 2001). In the absence of better infrastructure, the presentation mode for WAP alerts is typically basic text.

iMode users, on the other hand, can receive advertising e-mail alerts – similar to SMS – automatically through the 'always-on' feature. These alerts are currently also text-based, but link to CHTML pages. In Japan, the reception of optional initial push-based advertising services has not been positive, due to the proliferation of irritating junk mail (AFX Asia, 2001). As yet, market research in this area is limited.

Overall, current push services are very much in the lower left-hand quadrant of Figure 8.1. Until the availability of better hardware, software, and network infrastructure, services will remain basic. With faster, packet-based networks and more sophisticated devices, protocols, and software, richer push-based advertising is likely to emerge, pushing the possibilities into the top left-hand quadrant.

Wireless pull advertising

Any wireless platform with the capacity for browsing content can be used for pull advertising. WAP- and HTML-type platforms are the most widely used.

As in push-based modes, the essence to successful pull advertising is careful targeting to achieve relevance, positive response, and acceptance (Lot21, 2001). User perception is key, particularly with connection cost and limited screen space. One way that advertisers have approached this problem is by positioning wireless advertising as additional content. American Express, on the AvantGo network (via web clipping), offer a variety of value-added financial services content as a 'taster' to subscription material. In another example, a stock quote may carry with it a link to an online trading site, such as E*trade; the link does the promotional work of advertising, and is paid for like an ad, but the user perceives it as content. Unlike the wired Internet, successful wireless advertising is likely to be unobtrusive, add value to the user, and complement what the user is doing at the time, rather than interrupt it. As such, the line between advertising and service becomes blurred (Katz-Stone, 2001).

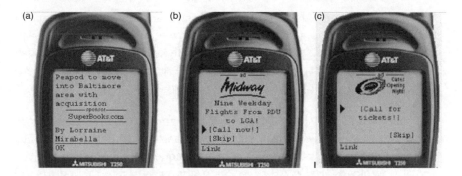

Figure 8.2 Types of WAP advertising (emulated WindWire ads): (a) simple text ad; (b) rich ad; and (c) interstitial ad

Figure 8.3 Advertisement on Palm VII PDA (emulated WindWire ad)

Various types of wireless pull ads have been created for mobile devices. Most WAP pull ads are simple in nature (bottom right of Figure 8.1), although HTML-type platforms offer higher richness. Figures 8.2 and 8.3 demonstrate several ads on an emulated WAP phone and Palm PDA. Figure 8.2 shows three key types of wireless ads, (a) to (c), as used in the WindWire First-to-Wireless advertising trial (Windwire, 2000). The types of adverts are as follows:

Simple text ads. These are the most basic level of WAP advertising. As the name suggests, they do not contain graphics or interactive response mechanisms. In Figure 8.2(a), SuperBooks is the sponsor of the displayed WAP page.

Rich ads. Building on the simple text ad, rich ads are defined primarily by their ability to be graphical and interactive, and can include WAP-compatible graphics, pull-down menus, click-though or call-through response mechanisms, and other interactive displays of advertiser messages. In Figure 8.2(b), the logo of the advertiser Midway Airlines is presented.

Interstitial ads. These rich ads use graphics and interactive text across the full device display – often referred to as 'splash' pages. These ads appear in the seconds between the user's request to view a site and the actual download of a WAP page. Such ads are usually cached on the device, and thereby do not increase download time. The advertisement in Figure 8.2(c) is for the Carolina Hurricanes, a US National Hockey League team. The advertisement is complete with a 'skip' option to bypass the ad. Interstitial ads can also be timed to close automatically after a specific period.

On the PDA advertising platform (and also iMode phones), the graphical capability is much greater, allowing rich ads and interstitials. Figure 8.3 shows an example of a splash page on the Palm VII. The advertisement is for Hoovers Online, a business information publisher, delivered on the 10Best Web site, a wireless-enabled publisher specializing in city guides and similar information. This ad is also available in colour.

One key issue in the delivery of wireless ads is the sheer volume of devices available. Without a standard that enables an ad to be deployed on a variety of sites and devices, the effectiveness of wireless advertising is diminished. The WAA, in association with players in wireless advertising such as WindWire, Lot21, SkyGo, and DoubleClick, have created standard ad formats for WAP and PDA devices (WAA, 2001). These standards are shown in Table 8.1 (shaded boxes are supported but not endorsed standards). With more advanced devices, these standards will need to be revisited.

Few companies are actually delivering advertising via WAP and PDA devices (Dawson, 2001). However, those firms that have conducted research trials have found encouraging results from consumers. WindWire, for example, surveyed 260 customers – ranging from users to non-users – using 2 million WAP and PDA ad impressions (105 individual ads and 22 ad campaigns). A total of six interactive agencies, seven wireless publishers, and 14 advertisers were involved. WindWire found that 51 per cent of users would be willing to view free ads, and 14 per cent would view ads unconditionally;

Table 8.1 Wireless advertising standards for WAP and PDAs

WAP ads		PDA ads	
Text ads	*Interstitials*	*Palm OS*	*Pocket PC OS*
One line, Fixed: 25 chars.	All standard sizes	150 × 24 pixels	215 × 34 pixels
Two lines, Fixed: 30 chars.	Ends after 4 s	150 × 32 pixels	215 × 46 pixels
One line, Marquee: 34 chars.	Skip option	Two lines of text	Two lines of text
Graphics plus text ads	*Graphics only ads*	*Palm OS*	*Pocket PC OS*
80 × 8 pixels + 15 chars. text	80 × 8 pixels	150 × 18 pixels	215 × 26 pixels
80 × 15 pixels + 15 chars. text	80 × 15 pixels	150 × 40 pixels	215 × 58 pixels
80 × 15 pixels + 30 char. text	80 × 20 pixels	One line of text	One line of text
80 × 20 pixels + 15/30 chars	80 × 31 pixels	Three lines of text	Three lines of text

some 91 per cent of users indicated that they would be very or somewhat influenced by wireless advertising (WindWire, 2000). Privacy was the chief concern among respondents (64 per cent).

Japan has experienced similarly positive responses to wireless pull advertising, using iMode. Interestingly, wireless advertising in Japan has more consumer appeal than advertising on the conventional Internet. Click-through rates for mobile banner ads during the summer of 2000 averaged 3.6 per cent, whilst those for wireless e-mail on iMode averaged 24.3 per cent. Click-through rates for online banner ads on desktop PCs in Japan often average no more than 0.5 or 0.6 per cent (Nakada, 2001).

A research model for wireless advertising

The discussion above has provided some insight into the nature of wireless advertising, particularly in terms of the wireless technological platform and basic applications of the medium. However, as yet, we have provided little conceptual discussion. The purpose of this section is to present a simple research model for further investigation. As such it consolidates the above discussion of mobile advertising

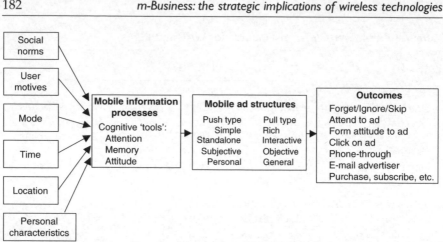

Figure 8.4　A research model for wireless interactive advertising

with other pertinent aspects of marketing, as supported by the relevant literature.

Figure 8.4 illustrates the basic components of the research model. As we move from the left-hand side of the diagram we encounter elements that encapsulate the individual's context and behaviour. Aspects such as social norms, user motives, user mode, time, location, and personal characteristics influence how the consumer processes information via cognitive tools. Such cognitive tools are employed on mobile ad structures, as controlled by the advertiser and varying along a number of dimensions. Finally, this interaction results in one or more outcomes. Let us examine the model in more detail.

Aspects influencing consumer information processing

As the above discussion has indicated, wireless advertising has considerable scope, based on the medium and the situation of the user. Potentially, interactive ads can be targeted cheaply and accurately, building one-to-one customer relationships, and reaching a large number of consumers (Meeker, 1997). However, there are a number of aspects that may be important in understanding how consumers process the ads, and, subsequently, how ads may be targeted to best effect. These include:

Social norms. Subjective norms towards mobile communications are likely to be influenced by normative beliefs that the individual attributes to significant others (friends, work colleagues, family members, and the like) with respect to adopting or continuing using the technology (Azjen, 1991). Further, it is now well recognized that mobile telephony plays a large role in social interactions. Since mobile phones are habitually used for social communication, the possibility of enhancing the effectiveness of messages via communities remains strong. Users often interact with their mobiles while with groups of friends 'showing off' any attractive new content (Puca, 2001). Therefore, virality is important – creating a message or experience that is so relevant and compelling that the consumer passes it on (Santello, 2001).

User motives. A motive is an inner desire to fulfil a need or want (Papacharissi and Rubin, 2000). When a user connects to the mobile network, there is some effort or involvement on the user's part, however minimal. For the Internet, more than 100 uses have been identified, falling largely into the categories of researching, communicating, surfing, and shopping (Rogers and Sheldon, 1999). As yet, the uses of the mobile medium are not as well known. The important point here is that because use is initiated with some specific goal in mind, an information-processing model must begin with these motives (Cannon *et al.*, 1998). These motives are likely to be antecedents to any ad processing that takes place once the motive is pursued. Ultimately, then, we expect that these motives influence consumer responses to mobile ads (Rogers and Thorson, 2000).

Mode. Mode is defined as the extent to which mobile use is goal-directed. This can be conceptualized as varying from 'telic' or serious goal-oriented activity, requiring more cognitive effort, to 'paratelic', low goal-directedness or playfulness, where users are more likely to be 'present-oriented', curious, and explorative (Katz and Aspden, 1997; Rogers and Thorson, 2000). On the Internet, evidence suggests that the latter are more likely to be receptive to online ads (Cho, 1998), and this is perhaps likely to also be the case for mobile advertising.

Time and *location.* An individual's behaviour and receptiveness to advertisement is likely to be influenced by their location, time of day, day of week, week of year, and so on. Individuals may have a routine that takes them to certain places at certain times, which may be pertinent for mobile advertising. If so, marketers can pinpoint location and attempt to provide content at the right time and point of need, which may, for example, influence impulse purchases (Kannan *et al.*, 2001). Feedback at the point of usage or purchase is also likely to be valuable in building a picture of time-space consumer behaviour.

Personal characteristics. The nature of the user, in terms of a plethora of personal characteristics such as age, education, socio-economic group, cultural background, and so on is likely important influence on how ads are processed. These aspects have already proven to be important influences on Internet use (OECD, 2001), and, as indicative evidence has shown above, elements such as user age are proving an important influence on mobile phone usage. The wireless medium has a number of useful means for building customer relationships. Ubiquitous interactivity can give the customer ever more control over what they see, read, and hear. Personalization of content is possible by tracking personal identity and capturing customer data; the ultimate goal is for the user to feel understood and simulating a one-to-one personal relationship.

Mobile information processes

Considerable literature has been developed in the area of information processing – the area that lies between inputs and outputs in humans (Thorson, 1989). Cognitive psychology posits 'mental' activities such as attending, attitude formation, memory storage, and decision-making in processing information (Neisser, 1967). Modern 'stage models' of advertising processing, such as McGuire (1978) and Preston (1982), build on these ideas: consumers gather information from ads they 'attend' to, comprehend that information, link it to what they already know, evaluate the information, form attitudes and intentions to purchase, and as a function of these processes, consumer behaviour is created. MacInnes and Jaworski (1989) provide a more broadly defined information processing theory of advertising where needs, motivation, ability, attention, cognitive and emotional processing, and attitude formation are important elements. Little work has been done on understanding how information is processed in the mobile medium, but the concepts espoused in these theories are likely to play an important role. In the research model, we take a basic stance; once we have determined the antecedents, we assume a number of information processing or 'cognitive' tools that the consumer uses to attend to the ads, remember them and develop attitudes based on them before making a response.

Mobile ad structures

Core to any wireless advertising campaign must be the issue of positioning the ads (or targeting). To maximize consumer response, advertisers must be able to target their audience to deliver a relevant, value-adding message that will make an impact on the individual. This will increase the likelihood of a positive response. However, the wireless Internet medium is limited in its ability to deliver a robust, complete message. Typically, users have access to more than one channel for information. A trade-off between reach and richness is, therefore, required (Wurster and Evans, 2000); the wireless medium is more useful for channel synergy and extending the brand of a company or product, as part of an overall marketing strategy.

In the mobile space, ad structures can vary along a number of dimensions, most of which were discussed earlier. Ads can be either be pushed to a device, where they pop-up on screen, or pulled by the user, typically via browsing. The ad can be small, simple, and text-based, or contain rich multimedia. The former is more suitable for the current generation of mobile phones and the latter for PDAs and future generations of mobile phones. Ads can be standalone, static content, or encourage a response from the user, such as clicking on a link to a WAP page or a phone-through. The information can be subjective or objective. Objective aspects include those of font size, image size, positioning, and colour. Such aspects have shown to be important in the context of Internet advertising studies (e.g. Li and Bukovac, 1999). On the other side of the coin, subjective aspects include more complex processing and interaction, such as playfulness, informativeness, effort in navigation, and relevance (Eighmey, 1997). Finally, ads can be generally or locally targeted. General ads can be aimed at market segments; personal ads, on the other hand, can be targeted at the individual, via customization.

Overall, the experience of a mobile ad should be consistent with the brand's value proposition, positioning, personality, and visual look and feel (Santello, 2001). Customer insights and surpassing expectations can be important in making the experience truly ownable by the brand. The most successful wireless advertisers promise to be those firms with an existing presence (Lot21, 2001).

Outcomes

Once the campaign has been designed and ads have been positioned, it is important to know the outcomes of advertising and how these can be measured. Rogers and Thorson (2000) provide a useful summary of responses to online ads, building on the measures of effectiveness for traditional media (e.g. Wells, 1997). Attention is one core aspect, most commonly measured by recognizing or recalling ad cues. In the worst-case scenario, the user may forget or ignore a mobile ad, or select the 'skip' option on mobile interstitials. If attention is gained, other outcomes are likely to follow. The attitude towards an ad, which we noted earlier as an important component of information processing models, is another important outcome. Forming the right attitude to an ad is a key aim of advertisers.

Based on the above outcomes, the consumer may perform an action. These may include clicking a link, phone-through, e-mailing the advertiser, or purchasing a product or service. Click-through allows consumers to respond to an advertiser without typing Web addresses, which is particularly difficult on small devices. Call-through is a similar response mechanism, which allows the user, in a single click, to be directed to an advertiser-sponsored phone number (such as a call centre). In a similar fashion, the consumer may also be inclined to send an e-mail or text message to the advertiser, based on an ad. Finally, a key outcome of advertising involves the user taking some specifically defined action in response to an ad. This is also known as 'performance-based' advertising. Actions include a sales transaction, subscription or online registration. The three most common ways to buy or measure wireless advertising are cost-per-action (CPA), cost-per-click (CPC), and cost per thousand (CPM) (Adflight, 2000).

Conclusions

The immediacy, interactivity, and mobility of wireless devices provide a novel platform for advertising. The personal and ubiquitous nature of devices means that interactivity can be provided anywhere. Advertising is potentially more measurable and traceable. Furthermore, technologies that are aware of the circumstances of the user can provide services in a productive, context-relevant way, deepening customer relationships. The convergence between marketing, CRM, and m-commerce represents a potentially powerful platform for wireless advertising.

Nonetheless, it is important to keep the consumer in mind; the key to success is the management of and delivery upon user expectations. Already, the wireless Internet has demonstrated the need for temperance; the wireless Internet is not an emulator of or replacement for the wired Internet, it is merely an additional, complementary channel for services. Further, aside from initial pilot investigations, it is not abundantly clear how consumers will respond to the idea of mobile advertising. In essence, wireless 'push' advertising creates clutter in an otherwise clean channel. Considerable further detailed research is needed to investigate the response of consumers, including questions such as:

What type of ad formats or executions could succeed given the sparseness in space and text interface?

How will consumers process ads in the wireless medium? What types of communication outcomes will be best generated in this medium? Exposure and attention to ads are almost guaranteed (given the small space) – does this mean it will be a more effective medium compared to others?

What privacy issues are involved in time- and space-context targeting? How can these issues be effectively overcome to open up the wireless channel for ads?

Currently, wireless advertising is embryonic and experimental – the majority of wireless advertising is SMS based. The next generation of devices and networks will be important in the evolution of wireless advertising; higher bandwidth will allow rich and integrated video, audio, and text. In addition, considerable effort is needed in building consumer acceptance, legislation for privacy and data protection, standardizing wireless ads, and creating pricing structures. If these conditions hold, wireless could provide the unprecedented platform for advertising that has been promised. Clearly, it is too early to tell, but future research aimed at examining these fundamental issues will help to further understand the implications of wireless advertising.

Acknowledgements

An earlier version of this chapter appeared as: Barnes, S.J. (2002). Wireless digital advertising: nature and implications. *International Journal of Advertising*, **21**, 1–22.

References

Adflight (2000). *Online Advertising 101: An Introduction to Advertising on the Web*. Belmont, CA: Adflight, Inc.

AFX Asia (2001). NTT DoCoMo seeks approval for push-type advertising info service on iMode. http://www.mformobile.com/main.asp?pk=12306, accessed 28 August 2001.

Azjen, I. (1991). The theory of planned behavior. *Organizational Behavior and Human Decision Processes*, **50**, 179–211.

Cannon, H., Richardson, T., and Yaprak, A. (1998). Toward a framework for evaluating Internet advertising effectiveness. *Proceedings of the Conference of the American Academy of Advertising*, Lexington, KY.

Cho, C.H. (1998). How advertising works on the WWW: modified elaboration likelihood model. *Proceedings of the Conference of the American Academy of Advertising*, Lexington, KY.

Dawson, K. (2001). Wireless advertising forecast optimistic. http://www.commwem.com/article/COM20010618S0010, accessed 15 August 2001.

Dutta, S. and Segev, A. (2001). Business transformation on the Internet. In S. Barnes and B. Hunt, eds, *E-Commerce and V-Business*. Oxford: Butterworth-Heinemann, pp. 5–22.

Eighmey, J. (1997). Profiling user responses to commercial Web sites. *Journal of Advertising Research*, **37**, 59–66.

Enpocket (2002). *Consumer Preferences for SMS Marketing in the UK*. London: Enpocket.

Enpocket (2003). *The Response Performance of SMS Advertising*. London: Enpocket.

Ericsson (2000). *Wireless Advertising*. Stockholm: Ericsson Ltd.

Frost & Sullivan (2001). *World Mobile Commerce Markets*. London: Frost & Sullivan Ltd.

Hamblen, M. (2000). Hello, this is a wireless ad. http://www.thestandard.com/article/display/0,1151,20104,00.html, accessed 28 March 2001.

IDC Research (2001). A billion users will drive e-commerce. http://www.nua.ie/surveys/index.cgi?f=VS&art_id=905356808&rel=true, accessed 28 May 2001.

Gluck, M. (2001). *Online Advertising Through 2006*. London: Jupiter Communications.

Kannan, P., Chang, A., and Whinston, A. (2001). Wireless commerce: marketing issues and possibilities. *Proceedings of the 34th Hawaii International Conference on System Sciences*, Maui, Hawaii, January.

Katz, J. and Aspden, P. (1997). Motivations for and barriers to Internet usage: results of a national public opinion survey. *Internet Research: Electronic Networking Applications and Policy*, **7**, 170–188.

Katz-Stone, A. (2001). Wireless revenue: ads can work. http://www. wirelessauthority.com.au/r/article/jsp/sid/445080, accessed 28 March 2001.

Kong, D. (2001). Wireless 101 for marketers: parts I to VII. http:// www.wirelessadwatch.com/, accessed 15 August 2001.

Li, H. and Bukovac, J.L. (1999). Cognitive impact of banner ad characteristics: an experimental study. *Journalism and Mass Communication Quarterly*, **76**, 341–353.

Lot21 (2001). *The Future of Wireless Marketing*. San Francisco: Lot21, Inc.

MacInnis, D.J. and Jaworksi, B.J. (1989). Information processing from advertisements: toward an integrative framework. *Journal of Marketing*, **53**, 1–23.

McGuire, W.J. (1978). An information-processing model of advertising effectiveness. In H.L. David and A.J. Silk, eds, *Behavioral and Management Science in Marketing*. New York: Ronald.

Meeker, M. (1997). *The Internet Advertising Report*. New York: Morgan Stanley.

Nakada, G. (2001). I-Mode romps. http://www2.marketwatch.com/ news/, accessed 5 March 2001.

Neisser, U. (1967). *Cognitive Psychology*. New York: Appleton-Century-Crofts.

OECD (2001). *Understanding the Digital Divide*. Paris: OECD Publications.

Ovum (2000). *Interactive Advertising: New Revenue Streams for Fixed and Mobile Operators*. London: Ovum.

Papacharissi, Z. and Rubin, A.M. (2000). Predictors of Internet use. *Journal of Broadcasting & Electronic Media*, **44**, 175–196.

Preston, I.L. (1982). The association model of advertising communication process. *Journal of Advertising*, **11**, 3–15.

Puca (2001). Booty call: how marketers can cross into wireless space. http://www.puca.ie/puc_0305.html, accessed 28 May 2001.

Quios (2000). *The Efficacy of Wireless Advertising: Industry Overview and Case Study*. London: Quios, Inc./Engage, Inc.

Rodgers, S. and Sheldon, K.M. (1999). The Web motivation inventory: a new way to characterize web users. *Proceedings of the Conference of the American Academy of Advertising*, Albuquerque, NM.

Rogers, S. and Thorson, E. (2000). The interactive advertising model: how users perceive and process online ads. *Journal of Interactive Advertising*, **1**, http://jiad.org/vol1/no1/rodgers/

Santello, P. (2001). *Direct Experience Marketing*. San Francisco: Lot21, Inc.

Thorson, E. (1989). Processing television commercials. In B. Derwin, L. Grossberg, B. O'Keefe, and E. Wartella, eds, *Rethinking Communication: Volume 2: Paradigm Exemplars*. Newbury Park: Sage.

WAA (Wireless Advertising Association) (2001). WAA Advertising Standards Initiative: draft standards, 6/26/01. http://www.waa-global.org/press/standards_press.html, accessed 28 August 2001.

Wells, W.D. (1997). *Measuring Advertising Effectiveness*. Mahwah, NJ: Lawrence Erlbaum.

WestCyber News (2001). WAP phone carriers and agencies begin trial ad service. http://www.mobilemediajapan.com/newsdesk/, accessed 23 August 2001.

Windwire (2000). *First-to-Wireless: Capabilities and Benefits of Wireless Marketing and Advertising Based on the First National Mobile Marketing Trial*. Morrisville, NC: Windwire, Inc.

Wurster, T. and Evans, P. (2000). *Blown to Bits*. Boston: Harvard University Press.

Chapter 9

Banking on wireless devices

Introduction

The Internet and the mobile phone – two technological advancements that have profoundly affected human behaviour in the last decade – have started to converge. The products of this association are data services for mobile communication. Using a variety of platforms, such services are being created to enable mobile devices to perform many of the activities of the traditional Internet, albeit in a format reduced for the limitations of mobile devices.

Some of the first commercial applications of the mobile Internet involve wireless or mobile (m-) banking. These developments build on earlier ideas of customer channel extension through telephone banking and online banking. m-Banking taps the mobile customer channel in an effort to further reduce costs and enhance customer relationships (Durlacher, 1999). In some emerging markets, such as China, where Internet penetration is low, m-banking offers the opportunity to break into the online banking channel for the first time (Datta *et al.*, 2001).

This chapter explores developments in m-banking in detail. This is a very new area both in terms of the mobile Internet platform and the banking applications that have been developed for it. Therefore, the chapter first provides some background to m-banking technologies and their capabilities, before exploring the strategic market for m-banking and aspects of the customer's acceptance of this new banking channel. The chapter then continues by analysing the strategic implications of m-banking for different markets, and the principal benefits of m-banking to the banking sector. Finally, the chapter concludes with a summary and conclusions on the future impact and likely success of m-banking.

Background to m-banking

By its nature, m-banking is a very new phenomenon. In order to better understand the subject of the analysis in this chapter, this section provides a brief outline of some of the salient m-banking platforms and services. The emphasis here is on a strategic understanding rather than a technical understanding of services.

Definition of m-banking

m-Banking can be defined as a channel whereby the customer interacts with a bank via a mobile device, such as a mobile phone or a PDA. The emphasis is on data communication, and in its strictest form m-banking does not include telephone banking, either in its traditional form of voice dial-up or through the form of dial-up to a service based on touch tone phones.

m-Banking platforms

The first applications of m-banking were based in Finland. As early as 1992 customers of Merita Nordbanken were able to make bill payments and check account balances using a mobile phone (based on GSM networks). More recently, m-banking applications have relied on the development of some key standards for wireless electronic services and expanded to global markets. In general, the platforms used have been WAP and SMS. However, in Japan, the iMode platform, based on cHTML (and more recently Java), has been the dominant platform for m-banking. Let us briefly examine each of these platforms in more detail.

WAP banking

WAP is a universal standard for bringing Internet-based content and advanced value-added services to wireless devices such as phones and PDAs. In order to integrate as seamlessly as possible with the Web, WAP sites are hosted on Web servers and use the same transmission protocol as Web sites, that is, HTTP (3G Lab, 2000). The most important difference between Web and WAP sites is the application environment. Whereas a Web site is coded mainly using HTML, WAP sites use WML, based on XML. WAP data flow between the Web server and a wireless device in both directions via a gateway that sits between the Internet and mobile networks. A wireless device will send a request for information to a server, and the server will respond by sending packets of data, which are formatted for display on a small screen by a piece of software in the wireless device called a microbrowser.

Mobile banking was one of the first transaction-enabled services to be provided using the WAP platform, partly based on research into consumer perceptions of these emerging services. Early WAP research by Nokia suggested m-banking was likely to become a primary application of choice for those users most interested in wireless data (Arthur Andersen and J.P. Morgan, 2000). Around 60 per cent of Europe's largest banks now offer m-banking services, with nearly all using the WAP platform (Epaynews, 2002). In some markets, such as Hong Kong, nearly all large banks – such as HSBC, Dao Heng Bank, Hang Seng Bank, and Citibank – offer sophisticated m-banking services (Maude *et al.*, 2000).

Figure 9.1 shows some screen shots of a banking customer interacting with a WAP phone. The customer logs into the m-banking WAP service to conduct a variety of m-banking activities. Typically these include:

- check the balance of their account (as shown in Figure 9.1a);
- transaction enquiry (as shown in Figure 9.1b);

Figure 9.1 Screen shots of m-banking activities: (a) checking an account balance; (b) checking transaction information; (c) finding product information

- view the last transactions made (usually three or five);
- check the status of a cheque number;
- transfer funds from one account to another (usually only for more advanced m-banks);
- request a transaction statement;
- request a cheque book;
- cancel a service request;
- check the status of service requests;
- change a password;
- pay a utility bill or credit card (usually only for more advanced m-banks);
- find account information, for example, the interest rate;
- find product information (as shown in Figure 9.1c);
- examine a branch listing.

In addition, more sophisticated m-banking platforms may also offer stock and securities trading and access to the foreign exchange market.

One of the key concerns with using WAP as a platform for m-banking is user acceptance. WAP adoption by consumers is both patchy and limited. As of July 2001, the use of WAP phones was disappointingly low; just 6 per cent of Finnish and US mobile phone users access the Internet using their phones, compared with only 10 per cent in the United Kingdom and 16 per cent in Germany (eMarketer, 2001). In Japan, the success of WAP services has been greatest, with 11.15 million subscribers to the EZWeb WAP service in September 2002 (Mobile Media Japan, 2002).

SMS banking

SMS allows text messages of up to 160 characters to be sent to and from mobile handsets via a store-and-forward system. Around 500 SMS billion messages were sent in 2001 (Frost and Sullivan, 2002). Although 95 per cent of this is based on person-to-person communication and voicemail, other services such as m-banking are growing in popularity. However, only around 10 per cent of Europe's biggest banks offer SMS banking (Epaynews, 2002). This is despite the fact that the majority of phones have SMS capability, but only 10 per cent have WAP capability (Datta *et al.*, 2001). In the United Kingdom, they include the Cooperative Bank, HSBC, and First Direct (West, 2002).

Using SMS, banks can offer a variety of services triggered by the customer sending a message to the customer service centre and receiving messages on their phone. By data input via the phone key-pad, customers can typically:

- check the balance of their account;
- check the status of a cheque number;
- transfer funds from one account to another;
- view the last transactions made (usually three to five);
- request a transaction statement;
- request a cheque book;
- change a password;
- pay a utility bill (usually only for more advanced m-banks).

For example, a customer could check their account balance by send-ing an SMS message <bank name><password><account number> (e.g. 'ABANK 1234 110001234') to the bank's customer service. The bank replies with a message containing the relevant account information.

Other platforms

Various other platforms are available or have been proposed for m-banking activities. These include the extremely successful iMode platform in Japan (Barnes and Huff, 2003), which had a phenomenal 35 million subscribers in September 2002 (Mobile Media Japan, 2002). Commission on subscriptions provide DoCoMo, iMode's par-ent company, with annual revenues of more than $3.4 billion per year (Nakada, 2001). The platform includes a number of m-banking facilities similar to those described above. These create approxi-mately 7 per cent of iMode's traffic (Funk, 2000). However, iMode is not restricted to Japan; services have been developed in parts of Europe and Asia, and are planned for the United States (see chapter 3) (Iwatani, 2002; Pikula, 2002).

On the PDA platform, a more sophisticated level of interaction can be facilitated due to the nature of the devices. Web clipping technol-ogy, such as AvantGo, allows popular PDA devices, such as Palm, Handspring, and Blackberry, to access dynamic and updated HTML content via a modem. Web clipping is used in combination with applications stored on the device. Figure 9.2 shows a service with access to share trading facilities on the RIM Blackberry PDA.

Aside from remote m-banking and financial services, numerous ideas exist for POS applications. For example, using short-range wireless technology in phones and terminals, such as Bluetooth,

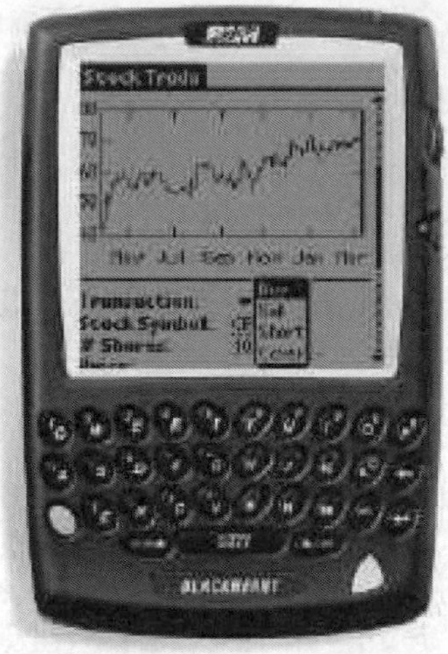

Figure 9.2 Stock trading on a mobile device

consumers are enabled to pay bills by connecting to the merchant's POS terminal, buy an item from a vending machine, pay a parking meter, and many other applications (Barnes, 2002b). The sum owed could, for example, be debited from a card or bank account or added to the phone bill.

The media richness of banking platforms

Clearly, the different platforms available for banking differ significantly in their ability to provide a rich, interactive banking experience, and to provide a high level of banking services. Although this is not a key focus of this chapter, it may be useful to make some comparisons between the various banking channels. One way to examine the capabilities of different banking platforms is in terms of information or media richness (Daft and Lengel, 1984, 1986). Communication richness is an invariant, objective property of communication media, and communication media differ in their ability to deal with and resolve ambiguity, converge multiple interpretations, and facilitate understanding. Media richness depends on four characteristics: availability of instant feedback (i.e. speed of communication); capacity of a medium to transmit multiple cues; personal focus; and richness of the language used (Daft and Lengel, 1984).

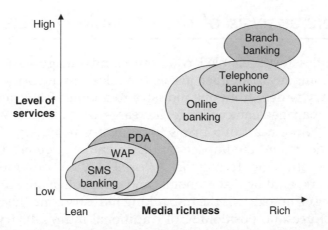

Figure 9.3 Media richness and level of services for banking platforms

Figure 9.3 demonstrates the differences in channel capabilities for banking. Generally speaking, face-to-face banking is ranked as the richest media due to its immediate feedback, multiple cues, natural language, and highly personal nature. At the other end of the spectrum, SMS banking is considered the leanest, since the bank customer cannot communicate with multiple cues and use of language variety is limited. In between we have telephone banking and online banking at the rich end, and WAP and PDA banking at the lean end. Alongside media richness, and related to channel capabilities, banking platforms vary considerably in the variety of services that they provide. Typically, branch banking provides the most services, followed by telephone and online banking. Mobile channels provide the least services, especially via SMS and WAP.

The implications of the differing channel capabilities are clear; mobile banking cannot yet replace traditional channels, where they exist, but provides a complementary channel optimized for mobility and personal access. This is similar to the lessons learnt in countries where online banking strategies have influenced bank branch closures and subsequent customer defection, such as the United Kingdom. Although superior to m-banking, even online banking cannot completely replace all aspects of branch or telephone banking; for example, online banking services typically have a telephone banking extension, and there are requirements for hardcopies of documents with signatures.

Strategic analysis of the m-banking sector

Opportunities for IT-enabled competitive advantage vary widely from one company to another, just as the rules and intensity of competition vary widely from one industry to another. The complexity of the IT management challenge increases considerably when IT penetrates to the heart of a firm's (or industry's) strategy. Thus, in order to understand the impacts of the m-banking we need a comprehensive strategic framework. Porter (1980) provides such as framework, arguing that economic and competitive pressures in an industry segment – such as the m-banking industry – are the result of five basic forces: (a) positioning of traditional intra-industry rivals; (b) threat of new entrants into the industry segment; (c) threat of substitute products or services; (d) bargaining power of buyers; and (e) bargaining power of suppliers.

This section aims to provide an analysis of the m-banking service sector from a strategic viewpoint. The purpose of this analysis is to provide some understanding of the key forces impacting on the ability of m-banking to succeed in provision of banking services. For illustration, we will draw some attention to the UK market, although the analysis can be reinterpreted for different markets. The results of such an analysis have distinct implications for practitioners. Figure 9.4 shows Porter's framework, highlighting some of the

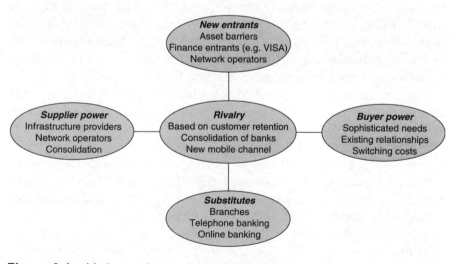

Figure 9.4   Market analysis – the m-banking sector

key elements for each of the five forces. As we can see, there is strength in each of the forces, emphasizing a very strong degree of competition in the provision of m-banking services. Let us examine the framework in more detail.

Rivalry

Banks are the major players in the provision of m-banking services. However, as Egg has shown in the online banking sector, the possibility exists for other players to enter the market offering specific products. Rivalry among banking providers differs from country to country. In the United Kingdom, the banking sector is competitive, with a large selection of banks in operation, although mergers and takeovers have essentially narrowed the market for the consumer. In this market, rivalry is not at a cut-throat level, with many banks focusing on building relationships with customers at a young age, rather than trying to lure established customers from competitors.

At present the m-banking sector is still embryonic; there are a handful of basic services being provided, with none of them being particularly well subscribed. Therefore, rivalry in this sector is currently low. A key issue from the banks perspective is the cost–benefit of providing new services. In the United States and Canada, some services have recently been switched off, including the Bank of America and the Bank of Montreal (Mobile Commerce World, 2002). One important lesson banks have learned from previous IT investments is the role of new channels in increasing the cost-per-customer (CPC). Previous investments in ATMs and telephone banking, for example, although with significantly lower average costs per transaction ($0.6 and $1.4 per month, respectively), were so convenient and popular that they induced high transaction volumes and thus higher costs per customer. During the period from 1985 to 1995, CPC increased by 39 per cent (Maude *et al.*, 2000). Since then, similar lessons have been learned from other IT investments such as online banking (Olazabal, 2002).

Alliances are likely to play an important role in developing m-banking services. No single industry alone has what it takes to establish m-banking. In the new mobile digital economy, consumer online services demand that diverse inputs must be combined to create and deliver value. The complementary assets and capabilities of mobile operators, banks, other financial services, infrastructure providers, software developers, network security specialists (such as

RSA Security or Verisign), and content providers are likely to play an important role, with banks and operators being at the core.

New entrants

Deregulation in the UK banking industry, and in many other developed economies, has meant increased competition for banks. This competition extends to the Internet. In the online banking sector, Internet-only banks such as Egg have challenged traditional 'bricks and mortar' banks. The same threat of new entrants in the m-banking sector is currently limited; customer perceptions of wireless services are not strong enough to build value in this area and there is a small customer base. Services based on WAP have received an apathetic response in many markets (eMarketer, 2001).

However, if customer perceptions strengthen, there is a real threat for a suitably positioned new entrant to build customer value. For example, a strong player in the payment market such as VISA may have some of the competencies and brand value to exploit customer demand. The possibility exists to build new, differentiated, value-added services for m-banking, such as inter-bank transfers, using VISA's market knowledge and expertise in secure transfers. A number of developments have already occurred in this area, including those of VISA and Nokia, Sonera, and France Telecom.

Perhaps an even larger threat exists from network operators; network operators already possess a powerful billing relationship with customers that can be leveraged to provide further financial services, thereby marginalizing banks. This model has already proved successful in Japan with iMode (Nakada, 2001).

Nonetheless, the barriers to entry for mobile banking are high. In order to enter the market a potential entrant must have an excellent knowledge of the market and consumer. Furthermore, the market is regulated and a banking license is required. The capital cost is also very high in terms of the IT spend.

Substitutes

A key threat to the viability of mobile banking is the availability of substitutes for the consumer. In the United Kingdom, the retail banking market has created an array of channels for banking services, including telephone, Internet, and branch banking. Any of these or other channels for interaction of customers with banking service providers must be considered substitutes for m-banking services. These services have built up large customer bases. At present, customers of alternative remote banking services need to be convinced of the added value of m-banking services in order to further build the customer base.

Buyers

Ultimately, customers hold the reins when it comes to channel selection. The banking sector tends to be based on a customer-centric model, where the emphasis is on managing customer relationships and meeting consumers' increasing levels of sophistication (Olazabal, 2002). Buyer power is increasing through deregulation and the provision of Internet banking solutions, although mergers and takeover are a further trend that has helped to reduce customer choice. The problems associated with branch closures in the United Kingdom and Ireland underline how customer behaviour is difficult to change. A variety of banks closed branches in remote areas, expecting customers to use other branches. In fact, customers protested, many affected customers moved to other banks, and a number of branches were reopened. Customers in banking are generally very localized, and although switching costs are not particularly high in the traditional market, banks exert extra costs by making such a move difficult (e.g. difficulty in closing bank accounts or transferring automatic payments).

On the other hand, if a service provides significant value to the consumer and is easy to learn and interact with, then, dependent on market conditions, customers tend to be drawn to it without a significant push from the banks. This is the lesson from telephone and online banking in the United Kingdom and similar markets. By the end of 2001, there were approximately 65 million online banking customers worldwide, of which 28 million were in Western Europe (eMarketer, 2002). Indeed, by 1999, 94 per cent of all European banks offered online banking, with similarly high rollout in other developed economies. In the United Kingdom, market conditions have helped the penetration of online banking: the cost of flat-rate Internet access is the lowest in Europe, and penetration rates are quite high (according to OFTEL, 2001, more than 40 per cent of people have access at home and 60 per cent at work). The same conditions do not yet exist for m-banking. While more than 70 per cent of the population have access to mobile phones, less than 10 per cent have access to WAP phones and many of these are light users (Forrester Research, 2002). The use of SMS on mobile phones is far more widespread, currently 186 billion per annum in Europe alone (Frost and Sullivan, 2002). However, while it is useful for push-based information (e.g. account balances), it does not offer the sophistication needed for key pull-based services (such as bank transfers), with more than 95 per cent of messages being used for person-to-person communication.

Suppliers

The main suppliers of wireless banking services are the parent banks. However, the companies involved in operating mobile services, providing the infrastructure (hardware and software) and developing content are also an important part of the value chain (Barnes, 2002a). Without appropriate competencies in the development and supply of mobile services, banks cannot reach the consumer market. The costs of obtaining such resources balanced against predicted demand is a major determinant of the banks' provision of m-banking services. This market is currently reasonably competitive, with a variety of operators competing for added value in the face of falling average revenue per customer. However, operators are beginning to leverage their infrastructure advantages in transport to enable movement along the value chain towards mobile services, delivery support, and market making (Barnett et al., 2000). Consolidation through merger and acquisition in this market is another trend. Such a move could strengthen the position of operators, who, ultimately, control the billing relationship with the customer and own the subscriber identity module (SIM). Competition in the provision of infrastructure also exists, but an oligopolistic practice among the well-known group of major players (such as Nokia, Lucent, Ericsson, Motorola, and Siemens) tends to keep prices high and stable (Kramer and Simpson, 1999).

Overall, the m-banking sector is very immature. It is subject to a variety of industry forces that each could play an important part in determining the future structure of competition. Customers, in accord with developing needs, are a powerful determinant of the success of this new banking channel over existing channels. As the sector develops, competition is likely to be created through rivalry from banks, potential new entrants from finance and telecommunications, and the increasingly powerful role of suppliers (especially operators and infrastructure providers).

The diffusion of m-banking services: lessons from technology acceptance theory

The adoption of wireless services is a complex phenomenon to predict. In some markets, wireless services have been met with apathy, such as in the United Kingdom, the United States, and Hong Kong (eMarketer, 2001). In other markets the success of services has been widespread; in Japan, iMode, launched in February 1999, has a subscriber growth rate of nearly 1 million per month (Mobile Media Japan, 2002). In this section, we try to unravel some of the complexities of technology adoption as applied to m-banking services. The aim is to gain some insight into the possible determinants of success or failure in m-banking adoption among consumers in different markets.

As a basis for the analysis, we will use technology acceptance theory. This theory examines the factors that influence the adoption, and diffusion throughout a social system, of new technologies or services such as m-banking. Figure 9.5 shows the basic model for our analysis. This is based on the work of a number of key theorists, particularly Rogers' (1995) research on technology diffusion, and Fishbein's and Ajzen's theory of planned behaviour (TPB) (Ajzen, 1991). Rogers (1995) suggests that there are five characteristics that are important in explaining the diffusion of innovations, such as m-banking: relative advantage, compatibility, complexity, trialability, and observability. TPB (Ajzen, 1991) posits that actual, voluntary use of a technology is determined by the individual's behavioural intention, which, in turn, is determined by the individual's attitude towards using the technology and subjective norms present in the individual's social environment. In recent years, Rogers' findings and those of TPB have been combined into a general theory of technology acceptance (Barnes and Huff, 2003; Karahanna *et al.*, 1999; Venkatesh and Davis, 2000), including additional factors to be considered along with Rogers's basic five, specifically image and trust.

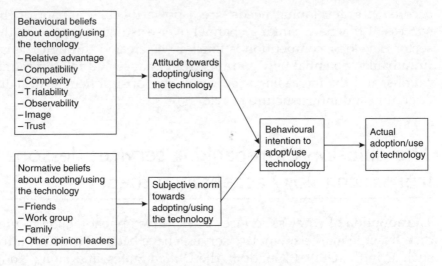

Figure 9.5 General model of technology acceptance (Barnes and Huff, 2003)

The model provides an interesting framework for analysing the potential of m-banking services. By assessing the attitudes and subjective norms associated with using the technology we are better equipped to examine aspects of the likely diffusion of m-banking in different markets.

Attitudes and behavioural beliefs towards adopting and using m-banking services

Figure 9.5 suggests that an individual's attitude toward adopting m-banking is determined by seven key characteristics. Let us examine each of these, in turn, before turning our attention to the subjective norms and normative beliefs associated with m-banking.

Relative advantage

A unique value proposition of m-banking services is that they are available 'on the move'. Many other aspects of the service, such as those of interactivity and remote access, are not unique. The relative advantage of m-banking is strongly dependent on existing banking services, electronic or otherwise. For example, m-banking in emerging markets – such as China or India – creates a principal source of relative advantage from the fact that it provides an access channel to many individuals for whom online banking was effectively inaccessible previously (Datta *et al.*, 2001). m-Banking provides the potential for low-cost access to remote banking services. However, in mature economies where remote banking services are more developed – such as the United Kingdom, the United States, the Netherlands, and Germany – the relative advantage of m-banking will be small. In these markets, managing the reach and richness of services is an important aspects of market positioning (Wurster and Evans, 2000). In this case, the consumer banking sector is already saturated with multiple banking channels such as telephone and online banking. Here, m-banking is a channel extension of previous services.

Complexity

The complexity of m-banking is dictated by the available technological infrastructure and usability of m-banking services. Present technological constraints include low power, difficulties in data input, and small screens for handheld devices (as well as high latency and low bandwidth for most existing services). In areas where traditional online access is available, the use of such devices is likely to be considered more complex and less satisfactory than online banking. This is certainly the case for many WAP banking services (Barnes and Vidgen, 2001). Areas without preconceptions about online services, such as developing markets, are likely to be more oriented towards new mobile services. However, it is easy to overestimate this constraint. Subscriptions to SMS services are growing at a phenomenal rate – over 30 per cent a month in many markets – despite the crude interface and display (Maude *et al.*, 2000). Similarly, the financial services provided on iMode, based on the familiar HTML and controlled by a simple and intuitive interface of menus and the command navigation button, have experienced rapid subscription growth (Funk, 2000).

Compatibility

The compatibility of m-banking with a customer's experiences and behaviour varies widely between different markets. A key part of this is the user's experience with other technologies and services. This could be in terms of the penetration of mobile phones, PCs, online banking, telephone banking, SMS, iMode, or WAP. For example, if the penetration of mobile phones is high and SMS is a popular service, this could provide a hook for SMS-based m-banking. Where communities are localized and the use of mobile telecommunications is low, the possibilities for m-banking may be more limited.

Trialability

Most market research conducted in developed economies shows that customers are understandably unwilling to pay for mobile services (e.g. account balances) that they already receive free through another channel; customers, however, are more willing to pay for mobile transaction services such as stock trading and for new pushed services that they could not receive any other way (Maude *et al.*, 2000). Nevertheless, advanced mobile banking services – even where they have been provided free of charge such as in Brazil and Turkey – have experienced relatively low customer take-up (Datta *et al.*, 2001). Part of the problem is encouraging users to try m-banking to demonstrate its value proposition, especially where observability is low. Trialability and pricing of services is likely to be an important part of establishing m-banking and building a critical mass.

Observability

The immediacy of m-banking services creates a relatively high level of observability. m-Banking usage is highly interactive, and, on some services, such as SMS and packet-switched network browsing, transactions are responded to almost instantly. Observability can be enhanced through individuals witnessing others using m-banking, especially since usage on mobile phones and PDAs often occurs 'out in the open'. However, unless contact is made with the user by the observer, it is often very difficult to establish what a device is being used for, and the security aspects of such services must also be observed. Observability is lessened somewhat due to the fact that some important aspects of the innovation, such as the network, are less visible. On the other hand, m-banking usage can form the basis for observable behaviour; for example, after a user checks a bank balance they may reorganize their finances, perhaps transferring money between accounts.

Image

Related to visibility, the image of device users is very important in some markets. Indeed, as the European, US, and Japanese markets have shown, many users select handsets and services on the basis of enhancing the individual's social status or image (Peter D. Hart, 2000).

It is possible that the same ideas of image may extend to ownership of m-banking services.

Trust

One of the major difficulties in the area of trust-building for m-banking services is security. In terms of technology, the mobile transaction channel is very new and needs to build a perception of security among consumers. Like the traditional Internet, this may take time, and may be easier for familiar services such as messaging. Human nature suggests that customers are more likely to trust traditional 'bricks-and-mortar' banks than those that they cannot see, weakening the possibility for building stand-alone m-banking services, even in emerging markets. However, the extension of trust for existing brands is a strong possibility, although the limited richness of mobile devices does make it more difficult to communicate the brand message to the consumer.

Normative beliefs and subjective norms towards adopting and using m-banking services

According to TPB, an individual's adoption of m-banking will be affected by subjective norms towards using the technology. These norms are determined by normative beliefs that the individual attributes to significant others (friends, work colleagues, family members, and the like) with respect to adopting or continuing to use the technology.

It is now well recognized that mobile telephony plays a large role in social interactions, and can be especially active amongst the younger age groups (Puca, 2001). This raises some concerns that there could be an age gap in the penetration of m-banking. Social norms will at least, in part, be related to those associated with existing infrastructure and service platforms, for example, via the penetration of mobile phones, SMS, iMode, or WAP. In many European countries, the usage of SMS is extraordinarily high among the youth and young adult segment. Similar patterns can be seen for iMode in Japan, where 70 per cent of subscribers are under 35. Individuals in this segment tend to be strongly influenced by their peers. Since mobile phones are habitually used for social communication, the possibility for passing on individual beliefs via communities tends to be strong. Users often interact with their mobiles while with groups of friends 'showing off' any attractive new content (Puca, 2001). Therefore, virality can be important – creating a message or experience that is so relevant and compelling that the consumer passes it on. The wireless digital space can be a fertile breeding ground for message and experience viruses (Santello, 2001).

Cultural and brand values can also form a basis for norms. For example, in Japan the cultural tendency toward group conformity – associated with iMode – suggests that a critical mass of usage leads to accelerated adoption and continued usage (Barnes and Huff, 2003). Further, loyalty towards certain brand identities in different countries – such as banks, network operators, and even handset providers – comprises a subjective norm that can further strengthen adoption and use of services such as m-banking.

Strategic implications for the banking sector

The success of mobile banking is likely to vary considerably from country to country. Basic services available in the more developed markets, such as in Europe, have met with some apathy. However, growth of mobile phones and m-banking services in some of the emerging markets could present a different picture. Figure 9.6 presents a simple strategic model for m-banking services, dependent on the

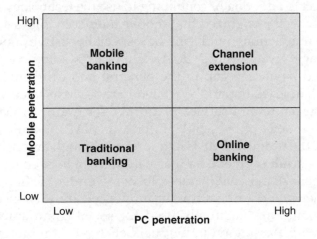

Figure 9.6   Strategic model for m-banking

relative level of penetration of data-enabled mobile phones and fixed-line PCs. This gives some indication of how m-banking is likely to be adopted in different markets.

Reliance on traditional banking in less developed markets

In countries where the impact of PCs and mobile phones is still very limited, such as sub-Saharan Africa, the impact of electronic banking has been negligible. Many consumers in these areas are remote and have low incomes. Many also do not possess a bank account at all. In these markets, there is a continuing reliance on traditional, over-the-counter banking. It would be foolhardy to suggest that mobile phones will be adopted just to allow access to m-banking. In reality, neither online banking nor m-banking are likely to succeed in these markets, except in possible niche areas.

Extension to multiple banking channels

At the other extreme, developed consumer markets have expanded to include a portfolio of multiple banking channels. In many developed markets where the penetration of PCs is high, online banking has become a complementary banking channel. In markets where mobile phone penetration is significant, the development of m-banking services will provide an additional banking channel, adding to the portfolio of banking channels available to the consumer. Such services may be more tailored and personalized for the consumer. In these markets, it is difficult, but not impossible, for m-banking to succeed as a stand-alone service, since it competes with powerful incumbent services and consumer perceptions. Examples of markets in this cell include the United Kingdom, Finland, the Netherlands, Hong Kong, and New Zealand.

Online banking as a complementary banking channel

Some developed markets have a high penetration of PCs but, as yet, a relatively low penetration of mobile phones. Such countries have developed online banking services, many of which have been successful channel extensions. However, the impact of mobile services, including m-banking, is likely to be low. The number of countries in this situation is relatively small and with the global expansion of mobile phone ownership, is becoming marginal or decreasing. Such markets include the United States and Canada.

Adoption of m-banking services in emerging and underdeveloped markets

In contrast, consumers and businesses in emerging mobile markets with limited wired networks are likely to find m-banking more attractive than their counterparts in developed markets. By 2005, when there will be more mobile phones in the world than televisions, fixed-line phones, or PCs, the lead of mobile devices in emerging markets will far exceed that of developed ones. There are some markets where m-banking could be positioned as a major channel for banking services, possibly even as a primary channel.

It is in emerging markets that mobile devices appear likely to beat PCs as the primary conduit of Internet services. Such markets include China, South Africa, and the Philippines. For many remote or low-income consumers, mobile handsets and the mobile Internet could for the first time provide access to basic banking and electronic payment services – a segment that banks find very difficult to serve cost-effectively. Some m-banking services can be built and rolled-out much more cheaply than traditional banking facilities (Datta *et al.*, 2001).

In addition, in developed economies where the wired Internet has not achieved high penetration, m-banking could provide a cost-effective banking channel for consumers. For example, countries such as Japan, Australia, Singapore, and Taiwan all have relatively low wired Internet penetration. Of these, Japan has the most developed market for m-banking services, providing an exemplar for successful services in similar economies.

Nonetheless, it is important to remember that the strategic role of m-banking services can develop over time. For example, developing countries could conceivably move from traditional banking to mobile or online banking. Similarly, countries with either mobile or online banking channels could move to a position of multiple electronic banking channels.

Strategic benefits of m-banking services

Figure 9.6 gives a simplistic framework for service positioning. Clearly, not all m-banking services are relevant to all markets; some are more suited to more financially sophisticated ones, others to the less developed. How much value an m-banking business can create depends largely on its relevance to a given market. Typically, in any market, a m-banking business can create value both directly and indirectly, as demonstrated in Figure 9.7.

Direct benefits

These benefits refer to cost savings and enhanced customer benefits:

1. *Cost savings*. Low-cost wireless solutions can provide for the first time an access channel for developing markets – such as China, India, and Nigeria. Typically, the high cost of elements such as ATMs and tellers have previously been prohibitive. However, the types of services provided need to be selective. Some services, for example, scan-based mobile POS payments – enabling customers to pay their bills – are feasible only in sophisticated mobile banking markets and in the affluent metropolitan areas of less-developed countries (e.g. Beijing, Mexico City, and Rio de Janeiro). Here, a few large organizations capture a high proportion of consumers' micro-spending and electronic payment would be familiar. In less-developed markets, such as India and Nigeria, a dial-up payments model would be more likely to work effectively (Datta *et al.*, 2001). Generally, the customer will bear most of the expense of accessing services. Access channels such as SMS are also a cheap and efficient way to reach individual consumers compared to telephone, letters, or email, especially for short, time-sensitive communications.

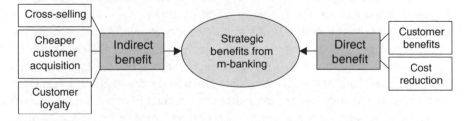

Figure 9.7 Strategic benefits from m-banking

2. *Customer benefits*. Mobile banking provides easy and convenient accessibility for the customer. Regardless of location, as long as it is connected to the network, a device can be used at any time to gain immediate access to services. Being able to manage finances in such a flexible way can also save costs for the consumer in terms of being able to avoid penalties or interest payments on accounts that are not paid, or to cancel automatic payments that are no longer required. Where other channels are available it provides the customer with a choice of access channels. In more sophisticated markets, services may be tailored to the consumer by mobile portals, although experience has shown that the value of these comes largely indirectly.

If a service provides obvious value to the bank consumer and is easy to use, then, dependent on the market situation, customers are likely to be drawn to it without a significant push from the banks. This is one lesson gleaned from both telephone and online banking in Europe and other developed markets. Where market conditions were right, e.g., in places where the cost of flat-rate Internet access was low and penetration rates were high, this provided an important spur to penetration of online banking.

Indirect benefits

These benefits include reduced churn rate, cheaper customer acquisition and cross-selling:

1. *Customer loyalty*. Looking deeper, the provision of m-banking services in more developed markets can help to strengthen customer loyalty and reduce customer churn. Customers in banking are generally very localized, Although they may move for specific services, customers tend to stay with banks for the longer term. Therefore, offering services that the consumer may demand reduces the tendency to move elsewhere.
2. *Cross-selling*. In the online banking market, cross-selling has become an important revenue stream. For example, Egg, the Internet-only bank in the United Kingdom, gains most of their revenue from selling insurance, financial services, and other products. Such ideas are likely to extend to the mobile channel, but are only likely to succeed in more sophisticated markets. In such markets, personalization and mobile portals may also be used.

3. *Cheaper customer acquisition.* As discussed above, the mobile channel opens the door to a large number of banking consumers in underdeveloped and emerging markets using, for example, low-cost banking and dial-up payments.

Summary and conclusions

The emergence of data communications on mobile phones and PDAs provides exciting possibilities for the growth of mobile banking services. Technical platforms such as SMS, iMode, and WAP have provided the potential channel for such services and hundreds of banks worldwide have begun to develop and use them (Wireless Financial Services News, 2002). This chapter has examined the nature of these services and their strategic potential as a viable channel for consumer banking. Porter's framework provided a useful structure to examine m-banking services, with a particular focus on the United Kingdom, the results of which provide some useful insights and implications for other markets. From a strategic perspective, the m-banking marketplace involves a complex interplay between banks and their existing banking channels, the consumer, and the suppliers of related competencies, technical infrastructure, and network services, some of which pose a potential threat in developing competing services.

The successful adoption and use of m-banking services is dependent on a number of factors, including existing banking channels, market conditions, and customer perceptions. We have examined the adoption of m-banking services using technology acceptance theory, underlining some of the key aspects of the technology and user behaviour that are likely to play a role in the future adoption. Whether the services become one of many channels or a primary channel for banking will significantly affect the potential return on m-banking investments by banks, which is estimated to be $254 million in 2002 in Western Europe alone (West, 2002). The issue of reach and richness of m-banking services is one that needs careful management (Wurster and Evans, 2000); even if a market existed, m-banking cannot replace fully the reach and richness of other banking platforms. However, as the above analysis has shown, it does provide significant potential for a channel to enable or extend electronic banking for consumers, depending on the prevailing market conditions.

The next 5 years will provide important evidence regarding the future potential of m-banking in existing and emerging markets. If industry analysts are correct, 2003 is likely to provide the inflection point for the growth of wireless data services (Jupiter Research, 2001), based on the penetration of cheaper data access via packet-switched networks (typically, early 3G implementations and GPRS) and the continued success of platforms such as SMS and its successor, MMS. Alongside, the growth of short-range wireless services, such as Bluetooth, could provide important complementary platforms for wireless financial services, especially electronic payment at POS.

Acknowledgements

An earlier version of this chapter appeared as: Barnes, S.J. and Corbitt, B. (2003). Wireless banking: concept and potential. *International Journal of Mobile Communications*, in press.

References

Ajzen, I. (1991). The theory of planned behavior. *Organizational Behavior and Human Decision Processes*, **50**, 179–211.

Arthur Andersen and JP. Morgan (2000). *Wireless Data: The World in Your Hand*. London: Arthur Andersen/JP Morgan.

Barnes, S.J. (2002a). Under the skin: short-range embedded wireless technology. *International Journal of Information Management*, **22**, 165–179.

Barnes, S.J. (2002b). The mobile commerce value chain: analysis and future developments. *International Journal of Information Management*, **22**, 91–108.

Barnes, S.J. and Huff, S. (2003). Rising sun: iMode and the wireless Internet. *Communications of the ACM*, in press.

Barnes, S.J. and Vidgen, R. (2001). Assessing the quality of WAP news sites: the WebQual/m method. *VISION: the Journal of Business Perspective*, **5**, 81–91.

Barnett, N., Hodges, S., and Wilshire, M. (2000). M-commerce: an operator's manual. *The McKinsey Quarterly*, No. 3, 163–173.

Daft, R.L. and Lengel, R.H. (1984). Information richness: a new approach to managerial behavior and organizational design. *Research in Organizational Behavior*, **6**, 191–233.

Daft, R.L. and Lengel, R.H. (1986). Organizational information requirements, media richness and structural design. *Management Science*, **32**, 5–15.

Datta, A., Pasa, M., and Schnitker, T. (2001). Could mobile banking go global? *McKinsey Quarterly*, No. 4, 71–80.

Durlacher (1999). Mobile commerce report. http://www.durlacher.com/research/, accessed 15 January 2000.

eMarketer (2001). Wireless Web growing around the world. http://www.nua.ie/surveys/index.cgi?f=VS&art_id=905357175&rel=true, accessed 10 September 2001.

eMarketer (2002). Banking on eEurope. http://www.emarketer.com/analysis/eeurope/20010801_europe.html?ref=dn, accessed 28 April 2002.

Epaynews (2002). European banks should offer SMS banking. http://www.epaynews.com/, accessed 3 May 2002.

Forrester Research (2002). Finance firms look to mobiles. http://www.nua.ie/surveys/index.cgi?f=VS&art_id=905357678&rel=true, accessed 23 April 2002.

Frost and Sullivan (2002). European SMS to double by 2006. http://www.nua.ie/surveys/index.cgi?f=VS&art_id=905357740&rel=true, accessed 29 April 2002b.

Funk, J. (2000). *The Internet Market: Lessons from Japan's iMode System.* Unpublished White Paper, Kobe University, Japan.

3G Lab (2000). *Your Pocket Guide to the Mobile Internet.* Cambridge: 3G Lab Limited.

Iwatani, Y. (2002). NTT DoCoMo may launch I-mode in US this year. http://story.news.yahoo.com/news?tmpl=story&u=/nm/20020319/wr_nm/telecoms_doco, accessed 22 March 2002.

Jupiter Research (2001). *Wireless Market to Take Off in 2003.* London: Jupiter MMXI.

Karahanna, E., Straub, D., and Chervany, N. (1999). Information technology adoption across time: a cross-sectional comparison of pre-adoption and post-adoption beliefs. *MIS Quarterly*, **23**, 183–207.

Kramer, R. and Simpson, B. (1999). *Wireless Wave II: The Data Wave Unplugged.* London: Goldman Sachs.

Maude, D., Raghunath, R., Sahay, A., and Sands, P. (2000). Banking on the device. *McKinsey Quarterly*, No. 3, 87–97.

Mobile Commerce World (2002). Wireless banking feels the pinch. http://www.mobilecommerceworld.com/, accessed 27 August 2002.

Mobile Media Japan (2002). Japanese mobile Internet users. http://www.mobilemediajapan.com/, accessed 15 October 2002.

Nakada, G. (2001). iMode romps. http://www2.marketwatch.com/news/, accessed 5 March 2001.

OFTEL (2001). *Consumers' Use of Mobile Telephony*. London: HMSO.

Olazabal, N.G. (2002). Banking: the IT paradox. *McKinsey Quarterly*, No. 1, 47–51.

Peter D. Hart (2000). *The Wireless Marketplace in 2000*. Washington DC: Peter D. Hart Research Associates.

Pikula, V. (2002). KPN starts tests with European iMode service. http://discuss.mobilemediajapan.com/stories/storyReader$3596, accessed 22 March 2002.

Porter, M. (1980). *Competitive Strategy*. New York: Free Press.

Puca (2001). Booty call: how marketers can cross into wireless space. http://www.puca.ie/puc_0305.html, accessed 28 May 2001.

Rogers, E. (1995). *Diffusion of Innovations*. New York: Free Press.

Santello, P. (2001). *Direct Experience Marketing*. San Francisco: Lot21 Inc.

Venkatesh, V. and Davis, F. (2000). A theoretical extension of the technology acceptance model: four longitudinal field studies. *Management Science*, **46**, 186–204.

West, L. (2002). Dial m for money. http://www.banktechnews.com/btn/articles/btjan01-3.shtml, accessed 5 May 2002.

Wireless Financial Services News (2002). *Wireless Banks and Brokers*. Newton, MA: Longwood Information.

Wurster, T. and Evans, P. (2000). *Blown to Bits*. Boston: Harvard University Press.

Chapter 10

Assessing the quality of WAP news sites – the WebQual/m approach

Co-author: Richard T. Vidgen

Introduction

The convergence of the Internet and wireless technologies is fuelling expectations of growth in wireless data services. One wireless data service platform for mobile computing is WAP – offering distinctive, small-screen Web offerings for handheld devices. Although developed specifically for the current technological environment, WAP is designed to be extensible enough to grow with the evolution of infrastructure. In the United Kingdom, WAP services have been widely available since autumn 1999 and the number of Web sites has grown considerably. However, as yet, there is little research into the qualities of a good WAP site, which we expect to be very different from that of traditional Web sites. With these issues in mind, we set about a programme of research aimed at assessing the general perception regarding current WAP offerings and developing an instrument capable of evaluating WAP sites. In earlier work, we reported the development of the WebQual instrument in the domain of traditional Internet the quality of Web sites (Barnes and Vidgen, 2000, 2001a,b). In this chapter, such prior research was used as the basis for an instrument adapted to take into account the WAP environment and mobile user context.

The implementation of WAP has provided both problems and opportunities (Barnett *et al.*, 2000; Logica, 2000). Nevertheless, it indicates an important starting point for the growth of wireless data services and is a de facto standard for wireless Internet services on digital mobile phones and other wireless terminals in many countries. With the provision of standards such as WAP, mobile telephony offers a potential platform for the penetration of a raft of services, including news, banking, gaming, shopping, and e-mail (Durlacher, 1999).

In this chapter, we report how the WebQual instrument has been adapted for WAP and subsequently tested in the domain of WAP news sites. In the next section, the development of WebQual/mobile (WebQual/m) is described. Subsequently, the collection of data for WAP news sites and the results of the data analysis are reported, including an examination of validity and reliability. Finally, a summary and some conclusions are provided along with an indication of the future development of WebQual/m.

The development of WebQual/m

Background to WebQual

WebQual (www.webqual.co.uk) is based on quality function deployment (QFD) – a 'structured and disciplined process that provides a means to identify and carry the voice of the customer through each stage of product and or service development and implementation' (Slabey, 1990). Applications of QFD start with capturing the 'voice of the customer' – the articulation of quality requirements using words that are meaningful to the customer. These qualities are then fed back to customers and form the basis of an evaluation of the quality of a product or service.

In the context of WebQual for traditional Web sites, users are asked to rate target sites against each of a range of qualities using a five- or seven-point scale. The users are also asked to rate each of the qualities for importance (again, using a five- or seven-point scale), which helps gain understanding about which qualities are considered by the user to be most important in any given situation. Although the qualities in WebQual are subjective (and quite rightly so), there is a significant amount of data analysis using quantitative techniques, for example, to conduct tests of the reliability of the WebQual instrument. In this sense, the approach bears some similarity to SERVQUAL (Parasuraman et al., 1988). WebQual also integrates some of the qualities of SERVQUAL (Barnes and Vidgen, 2001b), as well as those from the information quality and usability literature (Barnes and Vidgen, 2001a).

Reflections on the use of SERVQUAL

In the information systems domain, Pitt et al. (1995) argue that SERVQUAL is an appropriate instrument for assessing the service quality of the IS function. The basic SERVQUAL instrument was used with changes to the wording of the questions to make them relevant to the IS domain, where, for example, 'has modern equipment' became 'has up-to-date hardware and software'. Van Dyke et al. (1997) drew on experiences of the use of SERVQUAL in marketing to raise concerns about the application of SERVQUAL by Pitt et al. (1995) to IS service quality. Similar problems are identified in the work of Smith (1995). Chief among the issues raised were: concerns about the use of difference scores to measure the expectation gap; the unstable dimensionality of the SERVQUAL factors – different factors emerge in different applications of the instrument; and the difficulties of using a single instrument across industries. Van Dyke et al. (1997) note that 'the perception component of the difference score exhibits better reliability, convergent validity, and predictive validity than the perception-minus-expectation difference score' (p. 205) – taken in conjunction with concerns about the ambiguity of the perceptions construct they conclude that it is preferable to use a perceptions-only method.

The WebQual instrument uses Web site perceptions and importance to customer ratings equivalent to a two-column format, an approach that we believe captures the most important aspects of information and interaction quality whilst keeping the number of assessments that need to be made manageable. Although there is on-going debate about the SERVQUAL instrument, as Pitt *et al.* (1997) concluded, 'No good canvas is completed in a single attempt' and effort and hard work is needed to build a clear picture of IS service quality. The same is true of building a clear picture of Web quality, but the tried and tested ideas of service quality and its measurement provide a strong foundation for the development of WebQual.

WebQual/m is based on the WebQual research (Barnes and Vidgen, 2000, 2001a, b). Thus, the same techniques were used for evaluating WAP sites, although, of course, the focus of the evaluation is quite different. In particular, when creating the WebQual/m instrument, we were careful to provide a measure that is true to the current nature of the WAP service.

WebQual/m 1.0

As mentioned above, WebQual/m (infosys.bath.ac.uk/webqualm) takes onboard many of the key ideas of earlier research and applies them in the domain of WAP. Clearly, those that have used WAP will realize that it is a long way from the wireline Internet. Typically, early WAP services were criticized in terms of the restrictions of hardware (phone keypad, small screen, limited processing power), the slow speed, and its relatively high cost. However, the reality is that WAP provided a workable starting point at the embryonic stage of the mobile Internet (Barnes, 2002). The move to higher bandwidth services and richer technology (such as Java) is an evolutionary one, and WAP is the first in a series of steps that will be taken over in the next few years.

With these issues in mind, we were motivated to devise an instrument that reflected a reasonably accurate evaluation of the current nature of WAP services at the time of the research. The embryonic nature of WAP and the low penetration of the mobile Internet in the United Kingdom meant that it would not be possible to conduct focus groups in a meaningful way. Therefore, to this end, we re-evaluated the original WebQual instrument in light of the burgeoning professional literature on WAP (e.g. Arthur D. Little, 2000; Barnett *et al.*, 2000; Durlacher, 1999; Logica, 2000), aiming to produce a much-reduced questionnaire for the mobile context.

Table 10.1 WebQual/m questions and WebQual 3.0

No.	Description	WebQual 3.0
1	Has an appropriate appearance	Q3, Q4
2	Provides accurate information	Q7
3	Provides up-to-date information	Q9
4	Provides relevant information	Q10
5	Provides information which is easy to understand	Q11
6	Provides information at the right level of detail	Q12
7	Provides fast navigation to what I intend to find	*
8	Provides navigation which is free from errors	**
9	Is easy to find my way around and return to	Q1, Q2
10	Has content designed and selected for the mobility context	***
11	Has content which creates a sense of user community	Q17, Q18
12	Has interesting content	Q6

Notes: *This question did not appear in WebQual 3.0, but a similar version is found in WebQual 1.0
**This is a new question that combines some of the elements from question 7 of WebQual 3.0 with the key problem of navigational errors found in the professional literature (Durlacher, 1999)
***This is a new question aimed at assessing the suitability of content for the individual user's mobile context

The consensus in the literature is that WAP is currently constrained to largely informational use, and that in this situation quick and easy access to 'good' information for the user's mobile context are key. Thus, in devising WebQual/m, this meant that questions on the quality of information and navigation were most relevant – questions that have been a key part of WebQual since version 1.0. Table 10.1 presents the final WebQual/m instrument, indicating the relationship of the instrument with WebQual 3.0 (Barnes and Vidgen, 2001a). Notice that nine of the questions are clearly linked to WebQual. Of the remaining three, one is drawn from WebQual 1.0 and the others are new additions.

WebQual/m research design

In applying WebQual/m 1.0 we were keen to find a domain that would reflect the current capabilities of the WAP service. The rationale for choosing WAP news sites was clear: the provision of news is largely text-based or information intensive, and thereby provides a much higher suitability to task for the current technological infrastructure. In support of this, indicative research on usability found that news provision is one of the best-developed WAP services; the suitability to task of news (61.54 per cent) and booking a flight (42.14 per cent) are rated far in excess of shares prices (26.88 per cent) and e-mail (23.13 per cent) (*PC Magazine*, 2000).

The WAP news sites chosen were the BBC, *The Guardian*, and Excite (Reuters). These represent three of the first movers into WAP news services, all of which are large and well established in their traditional domains:

The British Broadcasting Corporation (BBC). The BBC has 2000 journalists in 55 bureaus throughout the world. BBC News reaches 70 per cent of the UK population (BBC News, 1999). BBC World Service radio broadcasts in 43 languages to every country in the world reaching 143 million listeners, whilst BBC World television broadcasts to 170 million households globally in 200 countries and territories. The BBC announced its WAP portal in December 1999, in partnership with Vodafone Airtouch, and it went live in January 2000. This provides an extension of its successful and award winning BBC News Online service on the Internet – launched in November 1997 and now the most popular news service outside the United States. Using the BBC's significant resources BBC News Online provides around 300 news stories per day, and an archive of nearly half a million stories. The WAP service provides a diverse range of news and information direct from BBC News to the WAP phone, including the top six general news stories as well as similar volumes of information on the City, sport, travel, TV, and science/ technology. When fully rolled-out the service will allow personalized news for the user, dependent on interests.

Excite Mobile. Excite was a first mover to the portal market. The Excite Mobile Portal is a free, fully featured WAP version of Excite's popular Web portal, which has around 1.2 million subscribers worldwide (Oreskovic, 2000). The WAP site, announced in January, was launched in February 2000 (Anderson, 2000). A broad range of optimized content is offered that utilizes Excite's personalization features – letting users pre-set category selections, for example, news, TV listings, sport, and stock prices. A total of 11 news categories are provided with content fed directly from Reuters.

The Guardian. The Guardian is one of United Kingdom's major broadsheet newspapers, with a market share of 16.65 per cent (as of July 2000). The Guardian Media Group also publishes *The Observer* – a Sunday newspaper – which enjoys a 15.47 per cent market share. *The Guardian* announced its WAP news service in January 2000, and it debuted shortly afterwards (Gapper, 2000). The WAP site is an extension of its award-winning network of Web sites – Guardian Unlimited. The News Unlimited WAP site includes the daily editorial content of *The Guardian* and *The Observer*, updated 24 hours a day. Other parts of the WAP site include Football, Cricket, Film, Books, Jobs, Education, and Work. All of these include breaking news, special reports, interactive features, and bulletin boards.

One key problem that we were aware of in designing the questionnaire was the small percentage of WAP phone ownership. Even though this is likely to be higher among the sample population (students) than elsewhere, it would still be very low. The solution was to make use of a WAP emulator, several of which are available for the Web. The WAP emulator chosen was acquired from www.gelon.net. In testing, this provided a quick, reliable, and relatively true-to-life imitation of a WAP phone. It also allowed for a variety of phone types, although the survey standardized on the Nokia 7110.

The WebQual/m questionnaire and associated survey was made available on the Internet and, as with WebQual, consisted of an opening instruction page that would then open a separate browser window containing the qualities to be assessed (Figure 10.1). In this case, the browser window used the Wapalizer from www.gelon.net and could be substituted by the user's own phone. The control panel allowed the user to switch the contents of the target window

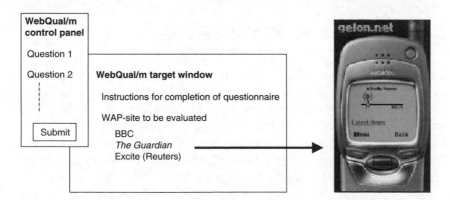

Figure 10.1   Internet-based WebQual/m questionnaire with emulated WAP site

between the instruction page and the target WAP site to be evaluated. This design allows the user to decide on the sequence of site evaluation and the order in which the questions are to be answered. For example, the user could decide to answer all questions for one site and then move on to the next site, or answer the same question for all three sites, or adopt a mixture of the two approaches.

In addition to the assessment of the sites using WebQual/m, we were also interested in gauging some general perceptions of WAP services. This would help to put WAP services in context and reinforce some of the ideas from the WAP literature. In particular, respondents were asked to select WAP services that appealed to them in the mobile context (binary answer from a list), and about perceptions of aspects of security, transactions, cost, and information access. These are discussed more fully in the next section as part of data analysis.

Data analysis

The data collected are summarized in Tables 10.2 and 10.3. Note that at this stage we have not presented any groupings of the questions to provide pertinent categories (this is discussed below). In all, we received 32 completed questionnaires and these formed the basis of the analysis discussed here. The body of responses came largely from students studying the MSc in Management (and variants) at the University of Bath. The questionnaire responses were received via e-mail, filtered to check for duplicates, and converted into a form usable in SPSS (a statistical software package) and Excel.

General perceptions of WAP services

To put the assessment of WAP in context, we were interested in examining students' general perceptions of these services. Figures 10.2 and 10.3 summarize these data. Older students, such as postgraduates, many of whom can expect a good job on graduation, are a large part of the target market for WAP data services (Peter D. Hart Research Associates, 2000). Although the sample is not large, it provides some indicative data in an area where research is scarce.

WAP can be used for many areas of service provision. Respondents were asked which areas they would be interested in

Table 10.2 Average and standard deviation information for the questionnaire

No.	Description	Importance		BBC		The Guardian		Excite (Reuters)	
		Avg.	St. Dev.	Avg.	St. Dev.	Avg.	St. Dev.	Avg.	St. Dev.
1	Has an appropriate appearance	3.34	1.21	3.03	0.90	3.47	0.72	3.38	1.07
2	Provides accurate information	3.75	0.98	3.34	1.00	3.41	0.87	3.72	0.68
3	Provides up-to-date information	4.13	1.39	3.47	1.24	3.38	1.24	3.88	1.07
4	Provides relevant information	3.59	0.84	3.13	0.94	3.38	0.71	3.63	0.71
5	Provides information which is easy to understand	3.53	1.11	3.38	1.24	3.47	0.72	3.97	0.90
6	Provides information at the right level of detail	3.53	0.72	3.38	0.98	3.41	0.84	3.78	0.83
7	Provides fast navigation to what I intend to find	3.91	1.53	2.78	1.10	3.03	1.03	3.28	0.96
8	Provides navigation which is free from errors	3.47	0.98	3.47	0.88	3.22	0.66	3.22	0.66
9	Is easy to find my way around and return to	3.53	1.16	3.28	1.57	3.41	0.84	3.91	1.15
10	Has content designed and selected for the mobility context	3.41	1.07	2.91	1.15	3.28	0.81	3.53	0.88
11	Has content which creates a sense of user community	3.72	0.85	3.59	0.87	3.59	0.76	4.22	0.91
12	Has interesting content	3.66	1.23	3.16	1.37	3.28	0.63	3.53	1.32

Table 10.3 Weighted scores and the WebQual Index (WQI)

No.	Description	Max. Score	BBC		The Guardian		Excite (Reuters)	
		Score	Wgt. Score	WQI	Wgt. Score	WQI	Wgt. Score	WQI
1	Has an appropriate appearance	16.72	10.25	0.61	11.47	0.69	11.22	0.67
2	Provides accurate information	18.75	13.25	0.71	13.50	0.72	14.25	0.76
3	Provides up-to-date information	20.63	15.72	0.76	15.00	0.73	17.00	0.82
4	Provides relevant information	17.97	11.50	0.64	12.00	0.67	13.00	0.72
5	Provides information which is easy to understand	17.66	11.91	0.67	12.66	0.72	14.19	0.80
6	Provides information at the right level of detail	17.66	12.44	0.70	12.19	0.69	13.72	0.78
7	Provides fast navigation to what I intend to find	19.53	11.16	0.57	12.88	0.66	13.91	0.71
8	Provides navigation which is free from errors	17.34	12.50	0.72	11.25	0.65	11.25	0.65
9	Is easy to find my way around and return to	17.66	11.22	0.64	12.34	0.70	13.97	0.79
10	Has content designed and selected for the mobility context	17.03	10.84	0.64	11.75	0.69	12.41	0.73
11	Has content which creates a sense of user community	18.59	13.91	0.75	13.81	0.74	16.06	0.86
12	Has interesting content	18.28	11.31	0.62	12.06	0.66	13.06	0.71
Total		217.81	146.00	0.67	150.91	0.69	164.03	0.75

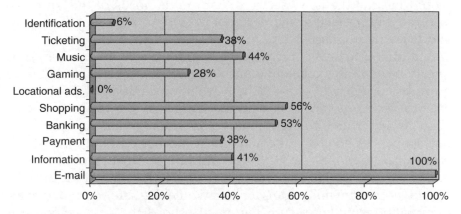

Figure 10.2 Preferred WAP services

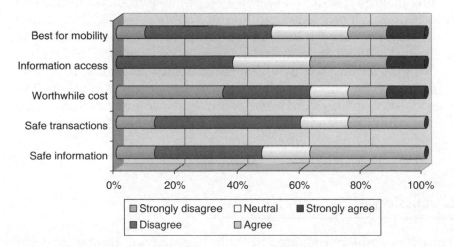

Figure 10.3 Perceptions of WAP services in general

using on a WAP phone – as drawn from the literature (e.g. Arthur D. Little, 2000; Durlacher, 1999; Logica, 2000). A list of 10 services was created for the questionnaire, along with a short description of each; respondents were then asked to select services they preferred in a binary fashion. Figure 10.2 shows the result. Responses ranged from 100 per cent for e-mail down to 0 per cent for location-based advertising. Using the phone as a means of identification and authentication also rated very low at 6 per cent. Of the remaining services, more than half would use banking or purchase online from mobile retailers. Music and information provision – the latter of which would include the news services that are the focus of this study – also rated quite well at more than 40 per cent, with mobile e-cash and ticketing not far behind.

In addition to areas of service, the survey also contained several questions relating to general perceptions of WAP. A sample of these questions is discussed here (others are not included in this chapter due to space restrictions). Specifically, respondents were asked about whether they thought that: WAP was superior to paper (e.g. newspapers or books) in the mobility context, access to information was satisfactory, the cost (£0.10 per minute at the time of survey – before packet-switched networks and pricing had been introduced) would be worthwhile, transactions would be safe, and personal information would be secure. Figure 10.3 shows the set of responses. As we can see, the views are somewhat mixed. There is quite strong feeling that the cost of WAP services is prohibitive and that safety of transactions is a problem. Both findings support issues discussed in the literature (Korpela, 2000; Logica, 2000; Manchester, 2000). There is similar feeling that paper-based information content is still endur-ing in the mobility context ('Best for mobility' in Figure 10.3), and quite mixed feelings about the safety of personal information. The most positive perceptions were about access to information – only 38 per cent of respondents felt that this was unsatisfactory, with no strong disagreement.

Discussion of the summary data

The data summary provided in Table 10.2 shows a number of items for discussion. The columns represent four data subsets based on the 32 responses: the importance rating for each question and the per question ratings for each of the three WAP news services (each of which was based on a scale of 1–5). Two summary items are shown in Table 10.2 for each question and subset: the arithmetic mean and standard deviation.

Referring to Table 10.2, there are notable patterns in the data. In terms of the importance ratings for the individual questions, these are quite high (with means of 3.34 and above), indicating some face validity to the components of the instrument. There are also some use-ful groupings to note. Those questions considered most important by the respondents – as indicated by means above the upper quartile of

3.73 – are associated with quick access to 'hard' information quality. Here we find, in descending order of importance, questions 3, 7, and 2. Such questions concern the accuracy and timeliness of information, as well as how quickly it can be found. Conversely, when examining those items considered least important – below the lower quartile of 3.52 – we find a quite different variety of questions. Specifically, questions 1, 10, and 8 are in ascending order of importance. This group revolves around softer issues such as aesthetics, the mobility context, and navigation errors. The remaining questions fall somewhere in between these two groupings and the median is 3.56.

The results suggest that there are a number of priorities demanded from WAP news service users. The results are rather pragmatic and underline quite strongly what we might expect from the current technology offerings; users are not so concerned with the look and feel of WAP sites, which is an obvious limitation, and have a pragmatic attitude towards the unreliability of navigation (as emphasized in the computer press: *PC Magazine*, 2000). Users appear to want immediate access to 'hard' information quality.

Turning to the results for the ratings of individual sites, we find some very stark contrasts. The unweighted averages in Table 10.2 show very clearly that the Excite UK (Reuters) WAP site ranks well above its two contemporaries. The site scored consistently higher than the other two sites for all but two items – questions 1 and 8 – where *The Guardian* and the BBC, respectively, each ranked top. The BBC and *The Guardian* are much closer in their scores, although the BBC ranks lower on all but three qualities.

WebQual indices

The weighted results shown in Table 10.3 serve to accentuate the differences indicated above in the direction of user priorities. Again, Excite (Reuters) appears to rank highest, although there is more competition in the rankings between the BBC and *The Guardian*. The total weighted scores give some indication of this.

Unfortunately, the weighted scores make it difficult to give a benchmark for the sites. One way to achieve this is to index the total weighted score for sites against the total possible score (i.e. the total importance multiplied by 5, the maximum rating for a site). To this end, Table 10.3 shows the maximum scores for each question and site. It also shows the WebQual Index (WQI) for each question and an overall WQI for each site (indicated in italics). Overall, Excite is benchmarked well above the other two WAP news services, with an overall WQI of 0.75. *The Guardian* follows with a WQI of 0.69, whilst the BBC is close behind at 0.67.

Perhaps more interesting is some conceptual assessment of how the WAP sites differ in quality. For this, we need to move beyond the scores and indices of individual questions towards a set of meaningful, reliable subgroupings. To this end, the next section derives a set of subcategories and applies them to the analysis.

Scale reliability and question subcategories

To verify the scale reliability of WebQual/m, a statistical reliability analysis was conducted using Cronbach's α (see Table 10.4). The test was conducted on the empirical data from each of the online news services. The resultant values of α averaged at 0.90, suggesting that the scale is quite reliable.

Furthermore, to better facilitate comparison between the WAP sites, reliability analysis was extended to a number of question subgroupings. Three categories were chosen, each with a further two subcategories. These can be explained as follows:

> *Information quality.* This includes hard aspects of quality such as accuracy and currency, as well as softer, more interpretive aspects such as relevancy, ease of understanding, and level of detail.
> *Site quality.* This includes aspects of the design of the site. WAP sites are not known for their aesthetic appeal, so the focus here is particularly geared around the appropriateness of appearance (format and style) and quality of navigation. Where sites are largely information-based, finding the right information quickly, easily, and without errors is of paramount importance.
> *User quality.* This refers to the WAP site's ability to meet the user's needs. These are the softest aspects of quality. Important questions include the suitability of content for the user's mobility context, whether such content is engaging for the recipient, and whether it is targeted at a suitable user community group in an appropriate way.

The average values of Cronbach's α are quite high for the categories, and two of them exceed 0.8. The user category has the smallest value

Table 10.4 Summary of reliability analysis for questionnaire categories

Subcategories	Questions	Average alpha
Information	Q2–Q6	0.86
Hard quality	Q2, Q3	
Soft quality	Q4–Q6	
Site	Q1, Q7–Q9	0.83
Navigation	Q7–Q9	
Appearance	Q1	
User	Q10–Q12	0.71
Empathy	Q11, Q12	
Mobility	Q10	

of α at 0.71, although this also has the smallest number of items – just three. Thus, it appears that we have a quite reliable set of question groupings. This provides a more solid foundation for assessment, based on a stronger recognition of the association between certain questions.

Site analysis using question subcategories

By utilizing the framework of categories examined in the previous section, we are able to build a profile of the qualities of an individual WAP site that is easily compared to its rivals. Thus, we are in a better position to examine why some sites fared better than others on the WebQual index. Figure 10.4 gives an example of how this can be achieved.

As a starting point, the data were summarized around the six questionnaire subcategories. Then, and similarly to the WebQual Index in Table 10.3, the total score for each category was indexed against the maximum score (based on the importance ratings for questions multiplied by 5). Figure 10.4 is the result, which rates the three Web sites using these criteria. Note that the scale has been restricted to values between 0.5 and 0.9 to allow for clearer comparison.

Figure 10.4 demonstrates very clearly that the Excite UK site stands head and shoulders above the two rivals. With the exception of site appearance, the indices for the subcategories for Excite make a clear circle around the other two sites, with mobility, user empathy, and

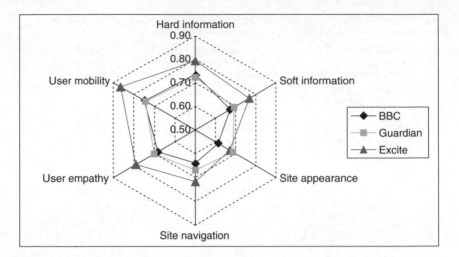

Figure 10.4 Radar chart of WebQual/m 1.0 subcategories for the three WAP sites

information quality rating particularly well. Other areas are less strong, in relative terms, although still ahead of other sites. In particular, site appearance ranked lowest for Excite, with other categories somewhere in between. Of the two other sites, *The Guardian* appears to be the stronger, although the scores are very close. *The Guardian* scores particularly well for site appearance, where it rates 8 points above the BBC and 2 points above Excite. However, the BBC edges slightly ahead of *The Guardian* for hard information quality and user mobility.

Overall, it is interesting to see that Excite is a clear leader in WAP news services. Excite is by far the most experienced player in providing electronic portal services, and this filters through strongly into its WAP offerings. The reliance on news feeds from Reuters, itself a key player in the electronic provision of news content, consolidates its position in the area of news services. The other players, both much less experienced in Web offerings, are some way behind. Of these two other players, *The Guardian*, whose competencies have traditionally relied on the written word, is slightly ahead of the BBC, which has traditionally focused on the spoken word (TV and radio). Nonetheless, as the technological infrastructure of the mobile Internet improves – particularly via faster, third-generation network transmission – players such as the BBC are likely to prosper due their multimedia experience.

The validity of WebQual/m

It is worth mentioning some points regarding the validity of the Webqual/m instrument. In developing the tool, we have paid attention to creating a scale that has some validity in measuring perceived WAP quality, drawing on an analysis and integration of the literature (for a full discussion, see Barnes and Vidgen, 2000, 2001a,b). Thus, in terms of the dependency on theoretical considerations, this adds a certain degree of validity. In addition, from the results of the research, the instrument does appear to have some usefulness in analysing the chosen Web sites; insofar as some aspects of Web site and service quality are observable, it appears to provide a reasonably valid indicator of criteria (IT Reference, 2000).

The questionnaire has purposefully been kept to a manageable size for data collection, and this is in line with similar work on service quality (Zeithaml *et al.*, 1993). However, whilst we believe that it captures many of the important items for WAP quality for news services, of course, it does not cover all possible items. Further relevant items may emerge as we further refine and test the instrument in different domains, and as the mobile technology infrastructure improves.

Summary and conclusions

WebQual/m is designed to give insight into customer perceptions of WAP site quality. The application of WebQual/m to news services shows that Excite is considerably ahead of its UK competitors, *The Guardian* and the BBC. Although analysis of the data suggests that the WebQual/m instrument has some reliability, further work is needed to refine and develop the instrument and to adapt it for different domains. Currently, information-intensive applications are dominant on the mobile Internet: there is not enough bandwidth for multimedia-intensive applications and there are very few services aimed at commercial activities such as product purchase. However, with improvements in transmission technology – such as third-generation networks – this is likely to change. Along with enhancements to service technologies like WAP (and implementation of those based on Java, for example), such advances will allow interactive and multimedia-rich applications for the mobile phone, resembling those of the Web today. Future revisions of WebQual/m will take more consideration of interactivity – based on an analysis of other important domains such as m-commerce and m-banking.

Furthermore, whilst WebQual/m has proved useful in giving a benchmark for WAP sites in terms of user perceptions, it says very little about how such issues translate into supply-side development. Based on QFD, future work will attempt to examine how the developer and organization can go about enhancing the perceptions of users, thereby increasing the value of the WebQual index. Such issues will require attention both to site and organizational aspects, the latter becoming particularly important as WAP sites become more interactive and commercial.

Acknowledgements

Thanks to Kenny Liu, an MSc student at the University of Bath, who was involved with the data collection for this research. An earlier version of this chapter appeared as: Barnes, S.J. and Vidgen, R.T. (2001). Assessing the quality of WAP news sites: the WebQual/m method. *VISION: the Journal of Business Perspective*, **5**, 81–91.

References

Anderson, T. (2000). Excite gets WAP portal. http://www.netimperative.com/, accessed 18 January 2000.

Arthur D. Little (2000). *Serving the Mobile Customer*. http://www.arthurdlittle.com/ebusiness/ebusiness.html, accessed 18 June 2000.

Barnes, S.J. (2002). The mobile commerce value chain: analysis and future developments. *International Journal of Information Management*, **22**, 91–108.

Barnes, S.J. and Vidgen, R.T. (2000). WebQual: an exploration of Web site quality. In *Proceedings of the Eighth European Conference on Information Systems*, Vienna, July.

Barnes, S.J. and Vidgen, R.T. (2001a). Assessing the quality of auction Web sites. In *Proceedings of the 34th Hawaii Internernational Conference on System Science*, Maui, Hawaii, January.

Barnes, S. and Vidgen, R, (2001b). An evaluation of cyberbookshops: the WebQual method. *International Journal of Electronic Commerce*, **6**, 11–30.

Barnett, N., Hodges, S., and Wilshire, M.J. (2000). M-commerce: an operator's manual. *The McKinsey Quarterly*, No. 3, 163–173.

BBC News (1999). Vodaphone and BBC News set the standards for future mobile and multimedia services. http://www.bbc.co.uk/, accessed 17 December 1999.

Durlacher (1999). *Mobile Commerce Report.* http://www.durlacher.com/, accessed 15 June 2000.

Gapper, J. (2000). Guardian aims for unlimited channels. http://www.netimperative.com/, accessed 27 January 2000.

IT Reference (2000). WAP sites for BT Cellnet mobile phones. http://www.itreference.com/WAP/, 15 August 2000.

Korpela, T. (1999). *White Paper of Sonera Security Foundation v. 1.0.* Helsinki: Sonera Solutions.

Logica (2000). *The Mobile Internet Challenge.* London: Logica Telecoms.

Manchester, P. (2000). Security fears are clouding growth forecasts. *Financial Times*, 5 July, p. v.

Oreskovic, A. (2000). Broadband or bust. http:// www. thestandard.com/article/article_print/1,1153,15755,00.html, accessed 13 June 2000.

Parasuraman, A., Zeithaml, V.A., and Berry, L. (1988). SERVQUAL: a multiple-item scale for measuring consumer perceptions of service quality. *Journal of Retailing*, **64**, 12–40.

PC Magazine (2000). Usability labs report – WAP services. September, pp. 134–147.

Peter D. Hart Research Associates (2000). *The Wireless Marketplace in 2000.* Washington DC: Peter D. Hart.

Pitt, L., Watson, R., and Kavan, C. (1995). Service quality: a measure of information systems effectiveness. *MIS Quarterly*, **19**, 173–187.

Pitt, L., Watson, R., and Kavan, C. (1997). Measuring information systems service quality: concerns for a complete canvas. *MIS Quarterly*, **21**, 209–221.

Slabey, R. (1990). QFD: a basic primer. Excerpts from the implementation manual for the three day QFD workshop. In *Transactions from the Second Symposium on Quality Function Deployment*, Novi, Michigan, June.

Smith, A.M. (1995). Measuring service quality: is SERVQUAL now redundant? *Journal of Marketing Management*, **11**, 257–276.

Van Dyke, T., Kappelman, L., and Prybutok, V. (1997). Measuring information systems service quality: concerns on the use of the SERVQUAL questionnaire. *MIS Quarterly*, **21**, 195–208.

Zeithaml, V.A., Berry, L., and Parasuraman, A. (1993). The nature and determinants of customer expectations of service. *Journal of the Academy of Marketing Science*, **21**, 1–12.

Index